ISSN 0044-2267

ZEITSCHRIFT FÜR ANGEWANDTE MATHEMATIK UND MECHANIK

INGENIEURWISSENSCHAFTLICHE FORSCHUNGSARBEITEN

BAND 61   1981   HEFT 9

AUS DEM INHALT

HAUPTAUFSÄTZE

THEORIA CUM PRAXI

AKADEMIE-VERLAG · BERLIN

ZAMM   Bd. 61   Nr. 9   S. 413—476   Berlin, September 1981

EVP 18,— M   34115

INHALT

Wir bitten, **Manuskriptsendungen** zweifach (Original und eine Kopie, sprachlich einwandfrei, Formeln mit Maschine oder in Druckschrift geschrieben) an folgende Anschrift zu richten:

Zeitschrift für Angewandte Mathematik und Mechanik
Institut für Mechanik der Akademie der Wissenschaften der DDR
DDR-1199 Berlin, Rudower Chaussee 5

Zu den Arbeiten, die als Hauptaufsätze bestimmt sind, ist auf gesondertem Blatt eine Zusammenfassung von 5 bis 10 Zeilen in englischer und (möglichst) deutscher und russischer Sprache beizufügen. Ausführliche Hinweise für die Autoren, um deren strikte Berücksichtigung gebeten wird, finden sich im Anschluß an das Inhaltsverzeichnis des Jahrganges **60** (1980).

Die Autoren erhalten von den Hauptaufsätzen 75, von den Kleinen Mitteilungen 25 Sonderdrucke ohne Berechnung, darüber hinaus weitere Sonderdrucke gegen Berechnung.

Der Verlag behält sich für alle Beiträge das Recht der Vervielfältigung, Verbreitung und Übersetzung vor.

Bestellungen sind zu richten

- **in der DDR** an den Postzeitungsvertrieb unter Angabe der Kundennummer des Bestellenden oder an den AKADEMIE-VERLAG, DDR-1080 Berlin, Leipziger Str. 3—4
- **im sozialistischen Ausland** an eine Buchhandlung für fremdsprachige Literatur oder an den zuständigen Postzeitungsvertrieb
- **in der BRD und Berlin(West)** an eine Buchhandlung oder an die Auslieferungsstelle KUNST UND WISSEN, Erich Bieber OHG, D-7000 Stuttgart 1, Wilhelmstraße 4—6
- **in den übrigen westeuropäischen Ländern** an eine Buchhandlung oder an die Auslieferungsstelle KUNST UND WISSEN, Erich Bieber GmbH, CH-8008 Zürich/Schweiz, Dufourstraße 51
- **im übrigen Ausland** an den Internationalen Buch- und Zeitschriftenhandel; den Buchexport, Volkseigener Außenhandelsbetrieb der Deutschen Demokratischen Republik, DDR-7010 Leipzig, Postfach 160; oder an den AKADEMIE-VERLAG, DDR-1080 Berlin, Leipziger Straße 3—4

ZEITSCHRIFT FÜR ANGEWANDTE MATHEMATIK UND MECHANIK
Herausgeber und Chefredakteur: Prof. Dr. Günter Schmidt. Redaktion: Dr. Winfried Heinrich, Dr. Horst Weinert, Dipl.-Math. Friedhild Dudel, Helga Rühl, Institut für Mechanik der Akademie der Wissenschaften der DDR. Verlag: Akademie-Verlag, DDR-1080 Berlin, Leipziger Straße 3—4; Fernruf: 2236221 oder 2236229; Telex-Nr.: 114420; Bank: Staatsbank der DDR, Berlin, Kto.-Nr. 6836-26-20712. Anschrift der Redaktion: Institut für Mechanik der Akademie der Wissenschaften, DDR-1199 Berlin, Rudower Chaussee 5; Fernruf: 6802841

Veröffentlicht unter der Lizenznummer 1282 des Presseamtes beim Vorsitzenden des Ministerrates der Deutschen Demokratischen Republik. Gesamtherstellung: VEB Druckerei „Thomas Müntzer", 5820 Bad Langensalza. Erscheinungsweise: Die Zeitschrift für Angewandte Mathematik und Mechanik erscheint monatlich. Die 12 Hefte eines Jahres einschließlich Tagungsheft bilden einen Band. Bezugspreis je Band 360,— M zuzüglich Versandspesen (Preis für die DDR 216,— M). Bezugspreis je Heft 30,— M (Preis für die DDR 18,— M). Bestellnummer dieses Heftes: 1009/61/9.

AN (EDV) 35937

ZAMM 61, 413–419 (1981)

W. SCHIEHLEN

Nichtlineare Bewegungsgleichungen großer Mehrkörpersysteme

Herrn Prof. Dr. F. WEIDENHAMMER zum 60. Geburtstag gewidmet.

Bei der dynamischen Untersuchung nichtlinearer mechanischer Systeme mit vielen Freiheitsgraden stellt bereits die Gewinnung der Bewegungsgleichungen ein wesentliches Problem dar. Es wird gezeigt, daß die Bewegungsgleichungen holonomer Mehrkörpersysteme durch die Anwendung des Jourdainschen Prinzips in einer einfacheren Form dargestellt werden können. Dies führt zu einem besseren mechanischen Verständnis symbolischer Gleichungen und zu kürzeren Rechenzeiten in numerischen Formalismen.

In the dynamical analysis of nonlinear mechanical systems with many degrees of freedom even the derivation of the equation of motion is an essential problem. It is shown that the equations of motion of holonomic multibody systems are strongly simplified by the application of Jourdain's principle. This results in a better understanding of symbolical equations of motion and a reduction of computation time.

При динамических исследованиях нелинейных механических систем со многими степенями свободы уже вывод уравнений движения представляет собой сложную задачу. Показывается, что уравнения движения голономных систем многих тел могут быть представлены в более простой Форме путём использования принципа Журдена. Это приводит к лучшему пониманию механического смысла символических уравнений и к сокращению машинного времени.

1. Einleitung

Das Problem der Aufstellung nichtlinearer Bewegungsgleichungen großer mechanischer Systeme ist zuerst in der Raumfahrttechnik aufgetaucht. Zur Untersuchung baumförmiger Satelliten entwickelten HOOKER und MARGULIES [1] im Jahre 1965 und ROBERSON und WITTENBURG [2] vom Jahre 1966 an Formalismen, die eine automatische numerische Erzeugung der Bewegungsgleichungen gestatten. Die Schwierigkeiten bei der praktischen Anwendung numerischer Formalismen dürfen jedoch nicht unterschätzt werden. So hat es z. B. zehn Jahre gedauert, bis der Formalismus von ROBERSON und WITTENBURG als Programmsystem von SCHWERTASSEK [3] dem Benutzer zur Verfügung gestellt wurde. Während ursprünglich nur Drehgelenke zwischen den einzelnen Teilkörpern eines Systems zugelassen waren, haben WITTENBURG [4], FRISCH [5] und ANDREWS und KESAVAN [6] auch allgemeine holonome Bindungen zugelassen und die auftretenden Zwangskräfte und -momente, z. B. mit dem D'ALEMBERTschen Prinzip, eliminiert. Mit dem nachlassenden Interesse an der Raumfahrttechnik und den wachsenden Anforderungen an die Fahrzeugtechnik und den Maschinenbau haben seit Mitte der Siebzigerjahre große Mehrkörpersysteme auch bei terrestrischen Aufgaben an Bedeutung gewonnen. Im Gegensatz zu den Satelliten besitzen die Systeme im Maschinenbau, z. B. bei Koppelgetrieben, häufig keine Baumstruktur. Die damit verbundenen kinematischen Schleifen lassen sich jedoch mit den numerischen Formalismen nicht direkt erfassen; sie erfordern den Einsatz LAGRANGEscher Multiplikatoren, die wiederum zu numerischen Instabilitäten führen können, siehe EIBER und SCHIEHLEN [7]. Parallel dazu setzte eine Rückbesinnung auf die klassischen Methoden der analytischen Mechanik ein, die keine Einschränkungen bezüglich holonomer Bindungen kennen. Da die LAGRANGEschen Gleichungen zweiter Art sich für große Systeme als sehr arbeitsintensiv und wenig rechnerfreundlich erweisen, wurden die NEWTON-EULERschen Gleichungen zusammen mit dem D'ALEMBERTschen Prinzip in eine computergerechte Matrizenform gebracht, SCHIEHLEN und KREUZER [8]. Es war sogar möglich, die Bewegungsgleichungen mit dem Digitalrechner in symbolischer Form aufzustellen. Das entsprechende Programmsystem ist seit 1978 für den Benutzer freigegeben, siehe z. B. KREUZER [9].

Die praktischen Anwendungen des Programmsystems für symbolische Bewegungsgleichungen haben gezeigt, daß lineare holonome Systeme mit bis zu fünfzig Freiheitsgraden ohne besonderen Aufwand untersucht werden können. Bei den nichtlinearen holonomen Systemen steigen Rechenzeit und Speicherplatzbedarf jedoch so stark an, daß maximal zehn Freiheitsgrade zulässig sind. Am Beispiel eines Industrieroboters hat dies KREUZER [10] gezeigt .Nachdem sowohl in der Kapazität der Digitalrechner als auch in der Leistungsfähigkeit symbolischer Programmiersprachen keine baldige Änderung zu erwarten ist, bleibt nur die Hoffnung auf eine Verbesserung der Ansätze der analytischen Mechanik. Hier ist nun eine Arbeit von KANE und LEVINSON [11] zu erwähnen, die am Beispiel eines Satelliten mit acht Freiheitsgraden und einer kinematischen Schleife im Innern sieben Methoden der analytischen Mechanik vergleicht und mit Hilfe sogenannter „partieller Geschwindigkeiten und Drehgeschwindigkeiten" zu überraschend einfachen Bewegungsgleichungen kommt. Die Methode der „partiellen Geschwindigkeiten" entpuppt sich bei näherer Betrachtung als eine wenig elegante Anwendung des JOURDAINschen Prinzips. KANE und LEVINSON [11] können deshalb auch keine Angaben über die Struktur der Bewegungsgleichungen und deren Zusammenhang mit den anderen Ergebnissen des Vergleichs machen.

In dieser Arbeit wird die allgemeine Struktur der Bewegungsgleichungen nichtlinearer holonomer Mehrkörpersysteme in Abhängigkeit von den verwendeten Koordinaten und Prinzipien untersucht. Es wird eine computergerechte Darstellung des JOURDAINschen Prinzips für holonome Systeme angegeben und der Zusammenhang mit anderen Ergebnissen aufgezeigt. Für die Praxis sind auch die linearen Bewegungsgleichungen bedeutsam. Diese werden ebenfalls diskutiert. Als Beispiel dient der von KANE und LEVINSON betrachtete Satellit.

2. Allgemeine Struktur nichtlinearer Bewegungsgleichungen

Das gegebene Mehrkörpersystem soll aus p starren Körpern K_i, jeweils mit der Masse m_i und dem 3×3-Trägheitstensor I_i, bezogen auf den Massenmittelpunkt C_i, $i = 1(1)\,p$, aufgebaut sein, Bild 1. Das System besitzt q holonome, rheonome Bindungen. Dann beträgt die Zahl seiner Freiheitsgrade

$$f = 6p - q \tag{1}$$

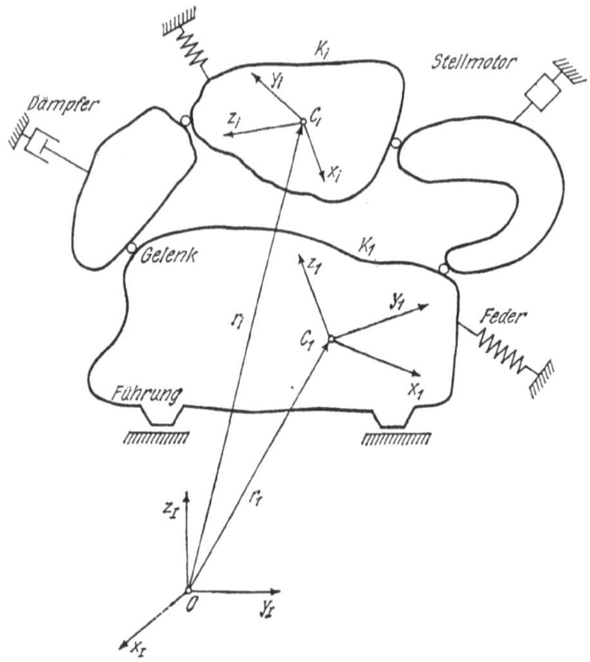

Bild 1. Mehrkörpersystem

und es genügen f verallgemeinerte Koordinaten zur eindeutigen Beschreibung seiner Lage, die zu einem $f \times 1$-Lagevektor $y(t)$ zusammengefaßt werden. Die Bindungen lassen sich in expliziter Form durch den 3×1-Ortsvektor

$$r_i = r_i(y, t)\,, \qquad i = 1(1)\,p\,, \tag{2}$$

des Massenmittelpunktes C_i und den 3×3-Drehtensor

$$S_i = S_i(y, t)\,, \qquad i = 1(1)\,p\,, \tag{3}$$

jeweils bezüglich des Inertialsystems x_I, y_I, z_I, beschreiben. Daraus erhält man durch Differenzieren die absoluten Geschwindigkeiten und Beschleunigungen für den i-ten Körper,

$$v_i = J_{Ti}(y, t)\,\dot{y} + \bar{v}_i(y, t)\,, \qquad \omega_i = J_{Ri}(y, t)\,\dot{y} + \bar{\omega}_i(y, t)\,, \tag{4), (5}$$

$$\dot{v}_i = J_{Ti}(y, t)\,\ddot{y} + K_{Ti}(y, \dot{y}, t)\,\dot{y} + \dot{\bar{v}}_i(y, t)\,, \qquad \dot{\omega}_i = J_{Ri}(y, t)\,\ddot{y} + K_{Ri}(y, \dot{y}, t)\,\dot{y} + \dot{\bar{\omega}}_i(y, t)\,, \tag{6), (7}$$

die neben den sonstigen Termen durch die $3 \times f$-JACOBI-Matrizen J_{Ti} und J_{Ri} gekennzeichnet werden. Die virtuelle Bewegung des i-ten Körpers, die im D'ALEMBERTschen Prinzip benötigt wird, lautet damit

$$\delta r_i = J_{Ti}(y, t)\,\delta y\,, \qquad \delta s_i = J_{Ri}(y, t)\,\delta y\,, \tag{8}$$

wobei δs_i der 3×1-Vektor der virtuellen infinitesimalen Drehung ist. Ebenso findet man aus (6) und (7) die virtuelle Geschwindigkeit des i-ten Körpers für das JOURDAINsche Prinzip als

$$\delta' v_i = J_{Ti}(y, t)\,\delta\dot{y}\,, \qquad \delta'\omega_i = J_{Ri}(y, t)\,\delta\dot{y}\,. \tag{9}$$

Führt man nun neben den verallgemeinerten Koordinaten noch einen $f \times 1$-Vektor von verallgemeinerten Geschwindigkeiten ein,

$$z(t) = H(y, t)\,\dot{y} + \bar{z}(y, t)\,, \tag{10}$$

wobei die reguläre $f \times f$-Matrix H und der $f \times 1$-Vektor $\bar{z}$ die kinematischen Beziehungen beschreiben, so folgt für die absoluten Beschleunigungen des i-ten Köpers

$$\dot{v}_i = L_{Ti}(y, t)\,\dot{z} + \dot{\bar{\bar{v}}}_i(y, z, t)\,, \qquad \dot{\omega}_i = L_{Ri}(y, t)\,\dot{z} + \dot{\bar{\bar{\omega}}}_i(y, z, t)\,, \tag{11), (12}$$

und die virtuellen Geschwindigkeiten lauten

$$\delta' v_i = L_{Ti}(y, t)\,\delta z\,, \qquad \delta'\omega_i = L_{Ri}(y, t)\,\delta z\,, \tag{13}$$

wobei die $3 \times f$-Matrizen L_{Ti} und L_{Ri} auftreten. Damit ist die Bewegung eines holonomen Mehrkörpersystems vollständig beschrieben.

2.1 NEWTON-EULERsche Gleichungen

Unter Berücksichtigung des Schnittprinzips gelten für jeden Körper K_i der Impulssatz

$$m_i\dot{v}_i = f_i^e + f_i^r\,, \qquad i = 1(1)\,p \tag{14}$$

und der Drallsatz

$$I_i\dot{\omega}_i + \tilde{\omega}_i I_i \omega_i = l_i^e + l_i^r\,, \qquad i = 1(1)\,p\,, \tag{15}$$

bezogen auf den Massenmittelpunkt C_i und angeschrieben im Inertialsystem oder im körperfesten System x_i, y_i, z_i. Die äußeren Kräfte und Momente sind in (14) und (15) in eingeprägte Kräfte f_i^e und Momente l_i^e und Reaktionskräfte f_i^r und -momente l_i^r aufgeteilt. Führt man nun die $6p \times 6p$-Massenmatrix

$$\operatorname{diag} M = \operatorname{diag}\,\{m_1E, m_2E, \dots, I_1, \dots, I_p\} \tag{16}$$

mit der 3×3 Einheitsmatrix E, die globalen $6p \times f$-Matrizen

$$\bar{J} = [J_{T_1}^T, J_{T_2}^T, \dots, J_{R_1}^T, \dots, J_{Rp}^T]^T\,, \qquad \bar{L} = [L_{T_1}^T, L_{T_2}^T, \dots, L_{R_1}^T, \dots, L_{Rp}^T]^T\,, \tag{17), (18}$$

und den $6p \times 1$-Vektor

$$\bar{q} = [f_1, f_2, \dots, l_1, \dots, l_p]^T \tag{19}$$

ein, so erhält man die NEWTON-EULERschen Gleichungen mit (6) und (7) in der Form

$$(\operatorname{diag} M)\,\bar{J}\ddot{y} + \bar{k}(y, \dot{y}, t) = \bar{q}^e(y, \dot{y}, t) + \bar{q}^r\,, \tag{20}$$

bzw. mit (11) und (12) als

$$(\operatorname{diag} M)\,\bar{L}\dot{z} + \bar{\bar{k}}(y, z, t) = \bar{q}^e(y, z, t) + \bar{q}^r\,. \tag{21}$$

Dabei ist $\bar{k}$, bzw. $\bar{\bar{k}}$, der $6p \times 1$-Vektor der Kreisel- und Zentrifugalkräfte und die eingeprägten Kräfte $\bar{q}^e$ werden lage- und geschwindigkeitsabhängig angenommen, so daß ein gewöhnliches Mehrkörpersystem vorliegt, siehe [12]. Die NEWTON-EULERschen Gleichungen enthalten die in den Bewegungsgleichungen unerwünschten Reaktionskräfte. Ihre Elimination erfolgt am elegantesten mit den Prinzipien der analytischen Mechanik.

2.2 D'ALEMBERTsches Prinzip

Nach dem D'ALEMBERTschen Prinzip in der LAGRANGEschen Fassung verschwindet in holonomen Mehrkörpersystemen die virtuelle Arbeit aller Zwangskräfte und -momente. Mit (8), (17) und (19) kann man dafür auch schreiben

$$\bar{J}^T\bar{q}^r = 0\,, \tag{22}$$

wobei $\delta y \neq 0$ beachtet wurde. Angewandt auf die NEWTON-EULERschen Gleichungen (20), verlangt das D'ALEMBERTsche Prinzip also die Linksmultiplikation der gesamten Gleichung mit $\bar{J}^T$ und es bleibt das bekannte Ergebnis

$$M(y, t)\,\ddot{y} + k(y, \dot{y}, t) = q(y, \dot{y}, t) \tag{23}$$

mit der symmetrischen, positiv definiten $f \times f$-Massenmatrix M und den $f \times 1$-Vektoren k und q der verallgemeinerten Kräfte. Die Bewegungsgleichungen (23) lassen sich auch mit den LAGRANGEschen Gleichungen zweiter Art gewinnen, doch ist der Arbeitsaufwand erheblich größer.

2.3 JOURDAINsches Prinzip

Das JOURDAINsche Prinzip wird im allgemeinen nur bei nichtholonomen Systemen angewandt, siehe [13], wenn das D'ALEMBERTsche Prinzip versagt. Das JOURDAINsche Prinzip besagt, daß bei nichtholonomen Mehrkörpersystemen die virtuelle Leistung aller Zwangskräfte und -momente verschwindet. Als allgemeineres Prinzip läßt sich aber das JOURDAINsche Prinzip auch auf holonome Mehrkörpersysteme anwenden. Dann gilt mit (9), (17) und (19) wiederum

$$\bar{J}^T\bar{q}^r = 0\,, \tag{24}$$

und man findet unter Berücksichtigung von (20) in unveränderter Weise die Bewegungsgleichungen (23).

Benützt man dagegen (13), (18) und (19), so bleibt

$$\bar{L}^T\bar{q}^r = 0\,, \tag{25}$$

wenn man $\delta' z \neq 0$ beachtet. Durch Linksmultiplikation der NEWTON-EULERschen Gleichungen (21) mit $\bar{L}^T$ erhält man jetzt die Bewegungsgleichungen

$$M^*(y, t)\,\dot{z} + k^*(y, z, t) = q^*(y, z, t) \tag{26}$$

mit der symmetrischen, positiv definiten $f \times f$-Massenmatrix M^* und den $f \times 1$-Vektoren k^* und q^* der verallgemeinerten Kräfte. Die Bewegungsgleichungen (26) lassen sich, wie KANE und LEVINSON [11] gezeigt haben, auch über die GIBBS-APPELLschen Gleichungen ausrechnen, wenn dies auch mit viel mehr Arbeit verbunden ist.

Die Bewegungsgleichungen (26) sind im Gegensatz zu (23) nicht alleine lösbar, sie müssen stets durch die kinematischen Differentialgleichungen (10) ergänzt werden.

2.4 Struktur- und Transformationseigenschaften

Die Struktur der abgeleiteten Bewegungsgleichungen wird deutlicher, wenn man sie im Zustandsraum darstellt. Mit dem $2f \times 1$-Zustandsvektor $x = [y^T z^T]^T$ lautet die Zustandsgleichung in beiden Fällen

$$B(x, t)\,\dot{x} = c(x, t)\,, \tag{27}$$

wobei aus (23) die $2f \times 2f$-Matrixfunktion

$$B(x, t) = \begin{bmatrix} E & 0 \\ 0 & M \end{bmatrix} \tag{28}$$

und die $2f \times 1$-Vektorfunktion

$$c(x, t) = \begin{bmatrix} \dot{y} \\ q - k \end{bmatrix} \tag{29}$$

oder aus (10) und (26) die Funktionen

$$B(x, t) = \begin{bmatrix} H & 0 \\ 0 & M^* \end{bmatrix}, \qquad c(x, t) = \begin{bmatrix} z - \bar{z} \\ q^* - k^* \end{bmatrix} \tag{30}$$

folgen. Durch Inversion von $B(x, t)$ erhält man aus (27) die Normalform nichtlinearer Zustandsgleichungen

$$\dot{x} = a(x, t) \tag{31}$$

mit der $2f \times 1$-Vektorfunktion $a(x, t)$. Für die numerische Integration erweist sich aber die Form (27) als günstiger.

Man erkennt nun durch den Vergleich von (28), (29) und (30), daß das D'ALEMBERTsche Prinzip die Zustandsgleichung in eine spezielle Form preßt, während das JOURDAINsche Prinzip einen größeren Spielraum in der Gestaltung erlaubt. Dies bedeutet, daß durch die Einführung und die geeignete Wahl von verallgemeinerten Geschwindigkeiten die Bewegungsgleichungen vereinfacht werden können.

Zwischen den Bewegungsgleichungen (23) und (26) besteht darüber hinaus ein enger Zusammenhang. Aus (6), (7), (11), (12) und (10) folgt unter Berücksichtigung von (17) und (18) die Beziehung

$$\bar{J} = \bar{L}H\,. \tag{32}$$

Damit umfaßt nach (22) und (25) das JOURDAINsche Prinzip in hinreichender Weise auch das D'ALEMBERTsche Prinzip. Weiterhin gilt für die Massenmatrix und die verallgemeinerten Kraftvektoren die Kongruenztransformation

$$M = H^T M^* H\,, \qquad k = H^T k^*\,, \qquad q = H^T q^*\,. \tag{33), (34}$$

Aus dem Impuls- und Drallsatz, (14) und (15), können nunmehr die Bedingungen für eine optimale Wahl der verallgemeinerten Geschwindigkeiten abgelesen werden. Im besonderen eignen sich als verallgemeinerte Geschwindigkeiten

1. eine oder mehrere Koordinaten der absoluten Geschwindigkeit (4) im Inertialsystem oder im körperfesten System, je nachdem, in welchem Koordinatensystem die eingeprägten Kräfte gegeben sind und

2. eine oder mehrere Koordinaten der absoluten Drehgeschwindigkeit (5) im körperfesten System (zeitinvarianter Trägheitstensor), wobei die Koordinaten, in denen eingeprägte Momente wirken, gegebenenfalls Vorrang haben.

In der Matrix H erscheinen unter anderem dann auch die Drehtensoren (3), was nach (33) zu erheblichen Vereinfachungen der Massenmatrix M^* im Vergleich zu M führt.

3. Lineare zeitinvariante Bewegungsgleichungen

Die Betrachtungen über lineare Bewegungsgleichungen sollen hier auf die Linearisierung der Gleichungen (23) bzw. (10) und (26) beschränkt werden. Dies reicht zur Untersuchung ihrer Struktur aus, nicht dagegen zur Entwicklung eines Programmsystems. Weiterhin soll vorausgesetzt werden, daß die Linearisierung auf zeitinvariante Koeffizientenmatrizen führt.

Die Linearisierung der Bewegungsgleichungen (23) bezüglich der Gleichgewichtslage ergibt die wohlbekannte Beziehung

$$M\ddot{y} + (D + G)\,\dot{y} + (K + N)\,y = h(t)\,, \tag{35}$$

siehe z. B. [8], während (10) und (26) auf

$$H\dot{y} = z - \bar{z}(t)\,, \qquad M^*\dot{z} + (D^* + G^*)\,z + Q^*y = h(t) \tag{36), (37}$$

führen, wobei Q^* eine $f \times f$-Matrix lageabhängiger Kräfte darstellt.

In der linearisierten Zustandsgleichung

$$\dot{x} = Ax + b(t) \tag{38}$$

mit dem unveränderten $2f \times 1$-Zustandsvektor $x = [y^T z^T]^T$ treten jetzt gemäß (35) die Matrizen

$$A = \begin{bmatrix} 0 & E \\ -M^{-1}(K+N) & -M^{-1}(D+G) \end{bmatrix}, \qquad b = \begin{bmatrix} 0 \\ M^{-1}h \end{bmatrix} \tag{39}$$

und nach (36), (37) die Matrizen

$$A = \begin{bmatrix} 0 & H^{-1} \\ -M^{*-1}Q^{*-1} & -M^{*-1}(D^* + G^*) \end{bmatrix}, \qquad b = \begin{bmatrix} -H^{-1}\bar{z} \\ M^{*-1}h^* \end{bmatrix} \tag{40}$$

auf. Auch hier zeigt sich durch den Vergleich von (39) und (40), daß das D'ALEMBERTsche Prinzip eine bestimmte Form der Zustandsgleichung erzwingt.

Der Zusammenhang zwischen den Bewegungsgleichungen (35) und (36) wird wiederum durch eine Kongruenztransformation vermittelt. Dabei gilt für die Matrizen M, D und G das Gesetz (33), während für die restlichen Matrizen die Beziehungen

$$K + N = H^T Q^*, \qquad h = H^T h^* - H^T M^* \bar{z} - H^T (D^* + G^*)\, \bar{z} \tag{41), (42}$$

gelten.

Die bei den linearen zeitinvarianten Systemen durch das JOURDAINsche Prinzip erzielbaren Vorteile sind allerdings nicht so gewichtig wie im nichtlinearen Fall. So muß zum einen die Inversion der Massenmatrix in (39) nur einmal, und nicht laufend bei jedem Integrationsschritt durchgeführt werden. Zum anderen können die lageabhängigen Kräfte, die in (37) durch Q^* gekennzeichnet sind, nicht mehr in einen konservativen (symmetrischen) Anteil und einen nichtkonservativen (unsymmetrischen) Anteil aufgespalten werden. Dadurch lassen sich die sehr weit entwickelten Stabilitätssätze der Mechanik, siehe MÜLLER [14], nicht mehr anwenden. Die qualitativen Aussagen über das Bewegungsverhalten werden so stark erschwert.

4. Satellitenbeispiel

Als Beispiel soll der von KANE und LEVINSON [11] betrachtete Satellit, Bild 2, aufgegriffen und mit der in dieser Arbeit entwickelten Methode untersucht werden.

Der Satellit besteht aus 4 starren Körpern und einem Massenpunkt, er besitzt 8 Freiheitsgrade. Seine Lage wird durch den 8×1-Vektor der verallgemeinerten Koordinaten

$$y = [q_1\, q_2\, q_3\, q_4\, q_5\, q_6\, q_7\, q_8]^T \tag{43}$$

beschrieben. Dabei sind q_1, q_2, q_3 die Kardanwinkel; weiterhin gilt $r_1 = [q_4 q_5 q_6]^T$, $q_7 = r$, $q_8 = \varphi$. Zur Vereinfachung der Bewegungsgleichungen wird zusätzlich der 8×1-Vektor der verallgemeinerten Geschwindigkeiten

$$z = [u_1\, u_2\, u_3\, u_4\, u_5\, u_6\, u_7\, u_8]^T \tag{44}$$

eingeführt. Zunächst werden die absoluten Drehgeschwindigkeitskoordinaten im körperfesten System als verallgemeinerte Geschwindigkeiten gewählt, $\omega_1 = J_{R1}[\dot{q}_1 \dot{q}_2 \dot{q}_3]^T = [u_1 u_2 u_3]^T$. Dann empfehlen sich die Absolutgeschwindig-

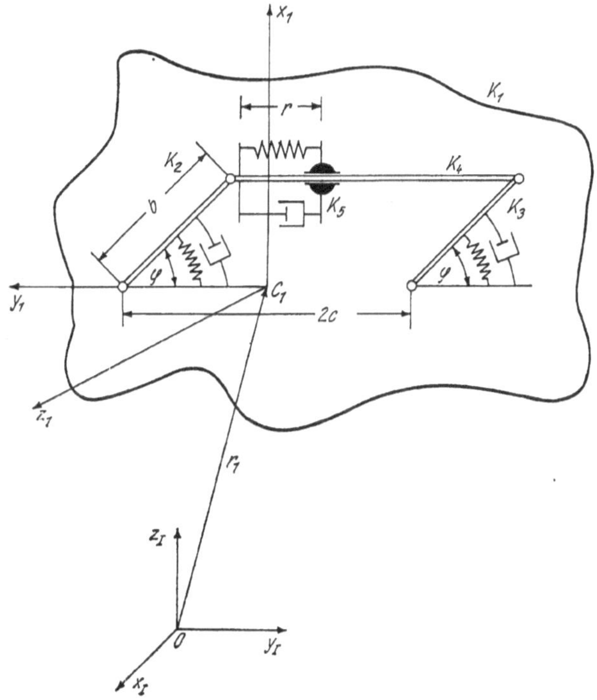

Bild 2. Satellit

keiten im körperfesten System $v_1 = S_1^T[\dot{q}_4 \dot{q}_5 \dot{q}_6]^T = [u_4 u_5 u_5]^T$, da die eingeprägten Kräfte in diesem System gegeben sind. Weiterhin ist $u_7 = b\dot{q}_8$ und $u_8 = b\dot{q}_8 s_8 - \dot{q}_7$. Damit ist bereits die 8×8-Matrix H der kinematischen Beziehungen gefunden:

$$H = \begin{bmatrix} c_2c_3 & s_3 & & & & & & \\ -c_2c_3 & c_3 & & & & & & \\ s_2 & & 1 & & & & & \\ & & & c_2c_3 & c_1s_3 + s_1s_2c_3 & s_1s_3 - c_1s_2c_3 & & \\ & & & -c_2s_3 & c_1c_3 - s_1s_2s_3 & s_1c_3 + c_1s_2s_3 & & \\ & & & s_2 & -s_1c_2 & c_1c_2 & & \\ & & & & & & & b \\ & & & & & & -1 & bs_8 \end{bmatrix}. \tag{45}$$

Als Abkürzungen wurden

$$s_i = \sin q_i \quad \text{und} \quad c_i = \cos q_i \tag{46}$$

verwendet. Alle nicht besetzten Elemente sind null.

Die fast immer regulären Matrizen H, M und M^* können in ausgearteten Lagen singulär werden. Tritt dieser Fall bei der Integration der Zustandsgleichung ein, so wird die Einführung eines zweiten Lagevektors mit anderen ausgearteten Lagen empfohlen. Mit den beiden nun zur Verfügung stehenden Bewegungsgleichungen wird die Integration dann abschnittsweise durchgeführt.

Berechnet man nun der Reihe nach die Beschleunigungen (11) und (12) und schreibt sie im körperfesten System x_1, y_1, z_1 an, dann findet man unmittelbar die 27×8-Matrix $\bar{L}$, die erwartungsgemäß nur schwach besetzt ist:

$$\bar{L} = \begin{bmatrix} & & & 1 & & & & \\ & & & & 1 & & & \\ & & & & & 1 & & \\ & & -\left(c - \frac{b}{2}c_8\right) & 1 & & & \frac{c_8}{2} & \\ & & \frac{b}{2}s_8 & & 1 & & \frac{s_8}{2} & \\ \left(c - \frac{b}{2}c_8\right) & -\frac{b}{2}s_8 & & & & 1 & & \\ & & \left(c + \frac{b}{2}c_8\right) & 1 & & & \frac{c_8}{2} & \\ & & \frac{b}{2}s_8 & & 1 & & \frac{s_8}{2} & \\ -\left(c + \frac{b}{2}c_8\right) & -\frac{b}{2}s_8 & & & & 1 & & \\ & & bc_8 & 1 & & & c_8 & \\ & & bs_8 & & 1 & & s_8 & \\ -bc_8 & -bs_8 & & & & 1 & & \\ & & -(c - bc_8 - q_7) & 1 & & & c_8 & \\ & & bs_8 & & 1 & & & 1 \\ (c - bc_8 - q_7) & -bs_8 & & & & 1 & & \\ 1 & & & & & & & \\ & 1 & & & & & & \\ & & 1 & & & & & \\ 1 & & & & & & & \\ & 1 & & & & & & \\ & & 1 & & & & \frac{1}{b} & \\ 1 & & & & & & & \\ & 1 & & & & & & \\ & & 1 & & & & \frac{1}{b} & \\ 1 & & & & & & & \\ & 1 & & & & & & \\ & & 1 & & & & & \end{bmatrix}. \tag{47}$$

Die weiteren Arbeiten beschränken sich nun auf das Ausschreiben der in (21) auftretenden Matrix (diag $\bar{M}$) und der Vektoren $\bar{\bar{k}}$ und $\bar{q}^e$ sowie die Linksmultiplikation von (21) mit $\bar{L}^T$. Diese Arbeiten können auch auf dem Digitalrechner symbolisch durchgeführt werden; ein entsprechendes Programmsystem befindet sich zur Zeit in Entwicklung.

Die Bewegungsgleichungen selbst sind von KANE und LEVINSON [11] bereits angegeben worden und sollen hier nicht wiederholt werden. Man kann sich leicht davon überzeugen, daß sie in die Form (26) gebracht werden können und alle oben genannten Eigenschaften besitzen.

5. Zusammenfassung

In dieser Arbeit wird eine rechnergerechte Methode zur Aufstellung symbolischer Bewegungsgleichungen von großen, gewöhnlichen Mehrkörpersystemen behandelt. Ausgehend von den NEWTON-EULERschen Gleichungen wird nach der Einführung verallgemeinerter Geschwindigkeiten das JOURDANsche Prinzip eingesetzt. Dadurch ist es möglich, die Bewegungsgleichungen in einer stark vereinfachten Form zu gewinnen. Die allgemeine Struktur dieser Bewegungsgleichungen wird aufgezeigt. Durch Kongruenztransformationen läßt sich der Zusammenhang mit den aus dem D'ALEMBERTschen Prinzip oder den LAGRANGEschen Gleichungen zweiter Art folgenden Bewegungsgleichungen herstellen. Die Linearisierung der Bewegungsgleichungen wird ebenfalls durchgeführt, wobei sich allerdings neben den genannten Vorteilen auch Nachteile, im besonderen für die Stabilitätsuntersuchung, ergeben. Die Methode wird an einem aus der Literatur bekannten Beispiel verdeutlicht. Insgesamt ist zu erwarten, daß die durch die Computertechnik gesetzten Grenzen für die Aufstellung symbolischer Bewegungsgleichungen großer nichtlinearer Mehrkörpersysteme weiter hinausgeschoben werden.

Literatur

1 HOOKER, W. W.; MARGULIES, G., The dynamical attitude equations for an n-body satellite, J. Astronaut. Sci. 12 (1965), S. 123 to 128.
2 ROBERSON, R. E.; WITTENBURG, J., A dynamical formalism for an arbitrary number of interconnected rigid bodies, with reference to the problem of satellite attitude control, Proc. Third Congr. Int. Fed. Auto. Control (IFAC), London (1966), S. 46D.1—46D. 9.
3 SCHWERTASSEK, R., Der Algorithmus von ROBERSON und WITTENBURG für Mehrkörpersysteme mit Baumstruktur, Deutsche Forschungs- und Versuchsanstalt für Luft- und Raumfahrt (DFVLR), IB 552-77/03, Oberpfaffenhofen 1977.
4 WITTENBURG, J., Dynamics of systems of rigid bodies, Teubner-Verlag, Stuttgart 1977.
5 FRISCH, H. P., A vector-dynamic development of the equations of motion for N-coupled rigid bodies and point masses, Nat. Aeron. Space Adm. (NASA), TN D-7767, Washington, D.C. 1974.
6 ANDREWS, G. C.; KESAVAN, H. K., Simulation of multibody systems using the vectornetwork model, in: MAGNUS, K. (Ed.), Dynamics of multibody systems, Springer-Verlag Berlin/Heidelberg/New York 1978, S. 1—13.
7 EIBER, A.; SCHIEHLEN, W., Dynanik geschlossener Gelenkketten. ZAMM, 60, T 38—T 40 (1980)
8 SCHIEHLEN, W.; KREUZER, E., Rechnergestütztes Aufstellen der Bewegungsgleichungen gewöhnlicher Mehrkörpersysteme, Ing.-Arch. 46 (1977) S. 185—194.
9 KREUZER, E., Symbolische Berechnung der Bewegungsgleichungen von Mehrkörpersystemen, VDI-Verlag, Düsseldorf 1979.
10 KREUZER, E., Dynamische Analyse offener Gelenkketten, ZAMM, 61, T 20 (1981)
11 KANE, T. R.; LEVINSON, D. A., Formulation of equations of motion for complex spacecraft, J. Guidance Control 3 (1980), S. 99 to 112.
12 SCHIEHLEN, W., Zur Klassifizierung von Mehrkörpersystemen, in: Deutsche Luft- und Raumfahrt, DLR-FB 75—32, Köln, DFVLR 1975, S. 254—269.
13 SCHIEHLEN, W. O.; KREUZER, E. J., Symbolic computerized derivation of equations of motion, in: MAGNUS, K. (Ed.), Dynamics of multibody systems, Springer-Verlag Berlin/Heidelberg/New York 1978, S. 290—305.
14 MÜLLER, P. C., Stabilität und Matrizen, Springer-Verlag Berlin/Heidelberg/New York 1977.

Eingereicht am 17. 10. 1980

Anschrift: Prof. Dr.-Ing. W. SCHIEHLEN, Institut B für Mechanik, Universität Stuttgart, Pfaffenwaldring 9, D-7000 Stuttgart 80, BRD

ZAMM 61, 421 – 433 (1981)

K. SHIRAKAWA

Dynamic Analysis of Cylindrical Shell under Hot Spot

Diese Arbeit beschäftigt sich mit der dynamischen Analyse einer zylindrischen Schale, die einer plötzlichen Änderung eines Temperaturfeldes durch örtliche Überhitzung zu einem Anfangszustand unterworfen wird. Unter Benutzung der Dirac-schen Delta-Funktion für die örtliche Überhitzung wird eine Lösung für die instationäre Gleichung des Wärmeflusses der Schale gegeben. Die dynamischen thermoelastischen zylindrischen Schalengleichungen, in denen sämtliche Trägheitsterme enthalten sind, werden mittels der Fourier- und Laplace-Transformation gelöst. Die Lösungen für ein quasistatisches Problem befähigen uns, die Singularitäten in den Wärmeverschiebungen und -spannungen zu diskutieren. Die Bedeutung der dynamischen Effekte in den praktischen Anwendungen wird durch einige numerische Beispiele ausgeführt. Die in dieser Studie benutzten Schalengleichungen entsprechen denen von Flügge und Donnell, die durch die Wärmeterme ergänzt wurden.

This paper is concerned with the dynamic analysis of a cylindrical shell subjected to a sudden change of temperature field by a hot spot at an initial state. Applying Dirac's delta function to the hot spot, the solutions are presented for the nonstationary equation of heat conduction for shell. The dynamic, thermoelastic cylindrical shell equation retaining all inertia terms are solved by employing Fourier and Laplace transforms. The solutions for a quasistatic problem enable us to discuss the singularities in thermal displacements and stresses. The importance of the dynamic effects in the practical applications is pointed out by showing some numerical examples. Shell equations employed in this study are those of Flügge and Donnell with the addition of thermal effects.

Данная работа посвящена динамическому анализу цилиндрической оболочки, которая посредством внезапного изменения температурного поля в связи с локальным перегревом приводится к начальному состоянию. Используя дельта-Функцию Дирака для локального перегрева строится решение нестационарного уравнения передачи тепла оболочки. Динамические термоупругие цилиндрические уравнения оболочки, содержащие все инерционные члены, решаются с помощью преобразований Фурье и Лапласа. Решения квазистатической задачи позволяют обсудить сингулярности в тепловых перемещениях и напряжениях. Значение динамических эффектов в приложениях показано на некоторых численных примерах. Использованные в данной работе уравнения оболочки соответствуют уравнениям Флюгге и Доннелла, дополненным тепловыми членами.

1. Introduction

The problem of thermal stresses is one of the important subjects in the fields of engineering applications such as welding, nuclear technology, mechanical, aerospace engineerings, etc. There have been a lot of books and studies on thermal stresses, but as for the shell structures, although the problem in the presence of heat source is rather important, there are few studies even for a cylindrical shell, which is one of the important elements of structures.

FLÜGGE and CONRAD [1], and JAHANSHAHI [2] obtained the solutions of shallow cylindrical shell subjected to hot spots. Recently, WILSON [3], SCHNELL and WARDENBACH [4] investigated the thermal stresses in cylindrical shells under arbitrary temperature distribution. All of these studies have been confined to the problems based on a prescribed temperature field or a quasistatic analysis. The dynamic, thermoelastic response of cylindrical shells was treated by MCQUILLEN and BRULL [5], who emphasized the importance of dynamic effects for all thickness within the realm of thin shell theory. As for the nonstationary problem of cylindrical shell, HUANG and WOO [6] considered the fundamental solution for a thermal shock on a finite orthotropic cylindrical shell, but the discussions were restricted only to the temperature distribution. The present author analyzed the quasistationary problem of cylindrical shell under hot spot [7] and recently extended this work to the study of dynamic problem of cylindrical shell under the local temperature distribution [8].

The purpose of this study is to develop the solution of the cylindrical shells under hot spots by a dynamic approach and to discuss in detail the thermal singularities from the view point of theoretical interest. The solutions obtained in this paper may be widely applied to practical problems, because the solutions due to a hot spot can be considered as GREEN's function. The equations of motion of cylindrical shell including thermal effects are introduced along lines identical to FLÜGGE [9] and DONNELL [10] formulation. The latter is based on some approximations of the former. The analysis is carried out by applying FOURIER and LAPLACE transforms.

The discussion on the thermal singularities is made for the solutions for a quasistatic problem. FLÜGGE and CONRAD [1] presented singular solutions based on the homogeneous equation, and JAHANSHAHI [2] those for the particular solutions, but their discussions were made for a shallow cylindrical shell. In the present paper transforming the solutions in terms of double FOURIER series, we obtain the solutions consisting of uniformly convergent series and the closed terms, which enable us to give the physical interpretation of the thermal singularities.

The hot spots treated in the present paper are related to the mean temperature rise and the temperature gradient through the thickness of shell. The former is designated as a plane hot spot and the latter as a bending hot spot.

2. Governing equations

Consider a thin circular cylindrical shell of radius a and thickness h, and take a coordinate system x, φ, z having its origin in the middle surface of the shell as shown in Figure 1 (a). Let us assume that the temperature $T_e(x, \varphi, z, t)$ at any given point of the shell may be expressed as

$$T_e(x, \varphi, z, t) = T(x, \varphi, t) - z\overline{T}(x, \varphi, t) , \tag{1}$$

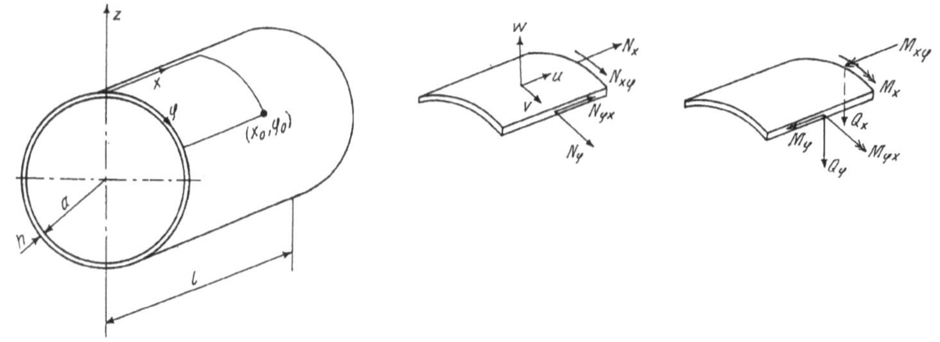

Fig. 1. Coordinate system, geometry and shell element with displacements, stress resultants and stress couples

where T is the mean temperature rise above a reference temperature and $\overline{T}$ the temperature gradient through the thickness of shell, respectively.

2.1 Equations of heat conduction

The nonstationary equation of heat conduction for shells was deduced by applying the variational principle to the problem of heat conduction by BOLOTIN [11], while MIZOGUCHI [12] introduced the same equation by the direct integration of the equation for the general three-dimensional heat conduction. Assume that the inner and outer ambient conditions of a shell are identical, then the nonstationary equations of heat conduction for an isotropic cylindrical shell in the absence of heat generation are separately given for T and $\overline{T}$ as follows:

$$\frac{\partial T}{\partial t} - \frac{\lambda_0}{c\varrho a^2}[T'' + T^{\cdot\cdot}] + \frac{2\varkappa}{c\varrho h} T = 0 \tag{2}$$

and

$$\frac{\partial \overline{T}}{\partial t} + \frac{\lambda_0}{c\varrho a^2}[\overline{T}'' + \overline{T}^{\cdot\cdot}] + \left[\frac{6k}{c\varrho h} + \frac{12\lambda_0}{c\varrho h^2}\right]\overline{T} = 0\,, \tag{3}$$

where t is time variable, c specific heat, ϱ density of material, k heat transfer coefficient, λ_0 coefficient of thermal conductivity, and $(\ )' = a(\ )/\partial x$ and $(\ )^{\cdot} = \partial(\ )/\partial\varphi$.

2.2 Equations of motion

The equations of motion including the thermal effects T and $\overline{T}$ are reduced along lines identical to the FLÜGGE type formulation [9] and the approximate equations to the DONNELL type formulation [10]. The detailed explanation is omitted and only the main point is described here.

Taking into account the relation (1) and following the FLÜGGE formulation, the stress resultants and the stress couples are related to the displacement components and the temperature rise as follows (see Figures 1 (b) and (c)):

$$N_\varphi = \underline{\frac{D}{a}(v^{\cdot} + w + \nu u')} + \frac{K}{a^3}(w + w^{\cdot\cdot}) - \underline{D(1+\nu)\alpha T}\,, \tag{4a}$$

$$N_x = \underline{\frac{D}{a}(u' + \nu v^{\cdot} + \nu w)} - \frac{K}{a^3}w'' - \underline{D(1+\nu)\alpha T} + \frac{K}{a}(1+\nu)\alpha\overline{T}\,, \tag{4b}$$

$$N_{\varphi x} = \underline{\frac{D}{a}\frac{1-\nu}{2}(u^{\cdot} + v')} + \frac{K}{a^3}\frac{1-\nu}{2}(u^{\cdot} + w'^{\cdot})\,, \tag{4c}$$

$$N_{x\varphi} = \underline{\frac{D}{a}\frac{1-\nu}{2}(u^{\cdot} + v')} + \frac{K}{a^3}\frac{1-\nu}{2}(v^{\cdot} - w'^{\cdot})\,, \tag{4d}$$

$$M_\varphi = \frac{K}{\underline{a^2}}(w + \underline{w^{\cdot\cdot} + \nu w''}) - \underline{K(1+\nu)\alpha\overline{T}}\,, \tag{4e}$$

$$M_x = \frac{K}{\underline{a^2}}(\underline{w'' + \nu w^{\cdot\cdot}} - u' - \nu v^{\cdot}) + \frac{K}{a}(1+\nu)\alpha T - \underline{K(1+\nu)\alpha\overline{T}}\,, \tag{4f}$$

$$M_{\varphi x} = \underline{\frac{K}{a^2}(1-\nu)}\left(\underline{w'^{\cdot}} + \frac{u^{\cdot}}{2} - \frac{v'}{2}\right), \qquad M_{x\varphi} = \underline{\frac{K}{a^2}(1-\nu)}\,(\underline{w'^{\cdot}} - v')\,, \tag{4g), (4h}$$

where ν is POISSON's ratio, E YOUNG's modulus of elasticity,

$$D = \frac{Eh}{1-\nu^2} \quad \text{and} \quad K = \frac{Eh^3}{12(1-\nu^2)}.$$

DONNELL's equations are derived if we take the underlined parts in the above equations. This convention will be adopted in the later discussions.

The equations of motion are written as

$$\underline{N_x' + N_{\varphi x}^{\cdot}} = \underline{a\varrho h \frac{\partial^2 u}{\partial t^2}}, \qquad \underline{aN_\varphi^{\cdot} + aN_{x\varphi}'} - M_\varphi^{\cdot} - M_{x\varphi}' = \underline{a^2\varrho h \frac{\partial^2 v}{\partial t^2}}, \tag{5a), (5b}$$

$$\underline{M_\varphi^{\cdot\cdot} + M_{x\varphi}^{\prime\cdot} + M_{\varphi x}^{\cdot\prime} + M_x'' + aN_\varphi} = \underline{-a^2\varrho h \frac{\partial^2 w}{\partial t^2}}. \tag{5c}$$

Substitution of (4a) through (4h) into (5a) through (5c) leads to the fundamental equations of motion with the thermal effects.

$$\underline{u'' + \frac{1-\nu}{2}u^{\cdot\cdot} + \frac{1+\nu}{2}v^{\prime\cdot} + \nu w'} + \varkappa\left[\frac{1-\nu}{2}u^{\cdot\cdot} - w''' + \frac{1-\nu}{2}w^{\prime\cdot\cdot}\right]$$

$$= \underline{\gamma^2 \frac{\partial^2 u}{\partial t^2} + (1+\nu)\,\alpha\alpha T'} - (1+\nu)\,\varkappa a^2\alpha \overline{T}', \tag{6a}$$

$$\underline{\frac{1+\nu}{2}u^{\prime\cdot} + v^{\cdot\cdot} + \frac{1-\nu}{2}v'' + w^{\cdot}} + \varkappa\left[\frac{3(1-\nu)}{2}v'' - \frac{3-\nu}{2}w^{\prime\prime\cdot}\right]$$

$$= \underline{\gamma^2 \frac{\partial^2 v}{\partial t^2} + (1+\nu)\,a\alpha T^{\cdot} - (1+\nu)\,\varkappa a^2\alpha \overline{T}}, \tag{6b}$$

$$\underline{\nu u' + v^{\cdot} + w} + \varkappa\left[\frac{1-\nu}{2}u^{\prime\cdot\cdot} - u''' - \frac{3-\nu}{2}v^{\prime\prime\cdot} + \underline{w'''' + 2w^{\prime\prime\cdot\cdot} + w^{\cdot\cdot\cdot\cdot}} + 2w^{\cdot\cdot} + w\right]$$

$$= \underline{-\gamma^2 \frac{\partial^2 w}{\partial t^2} + (1+\nu)\,a\alpha T} - (1+\nu)\,\varkappa a\alpha T'' + (1+\nu)\,\varkappa a^2\alpha \overline{T}'' + (1+\nu)\,\varkappa a^2\alpha \overline{T}^{\cdot\cdot}, \tag{6c}$$

where

$$\varkappa = h^2/12a^2, \qquad \gamma^2 = \varrho a^2(1-\nu^2)/E.$$

3. Theoretical analysis

Consider a cylindrical shell of finite length l, which is heated at time $t = 0$ by a plane and/or a bending hot spot at $(x, \varphi) = (x_0, \varphi_0)$ instantly and immediately afterwards cooled naturally (Figure 1 (a)). The analysis is carried out for a cylindrical shell with simply supported ends; for the thermal boundary conditions

$$T(0, \varphi, t) = T(l, \varphi, t) = 0, \tag{7}$$

$$\overline{T}(0, \varphi, t) = \overline{T}(l, \varphi, t) = 0 \tag{8}$$

and for the boundary conditions of edges

$$w(0, \varphi, t) = v(0, \varphi, t) = N_x(0, \varphi, t) = M_x(0, \varphi, t) = 0, \tag{9a}$$

$$w(l, \varphi, t) = v(l, \varphi, t) = N_x(l, \varphi, t) = M_x(l, \varphi, t) = 0. \tag{9b}$$

3.1 Solutions of heat conduction

Denote the intensity of a plane hot spot by Q and that of a bending one by $\overline{Q}$, then we can express the temperature distribution at time $t = 0$ by use of the DIRAC's delta function [13]

$$T(x, \varphi, t; x_0, \varphi_0, 0) = Q\delta(x - x_0)\,\delta(\varphi - \varphi_0)/a, \tag{10}$$

$$\overline{T}(x, \varphi, t; x_0, \varphi_0, 0) = \overline{Q}\delta(x - x_0)\,\delta(\varphi - \varphi_0)/ah. \tag{11}$$

The present author obtained the solution of equation (2) by applying the separation of variables [7], while HUANG and Woo by using the LAPLACE transform [6]. Taking into account the boundary conditions (7), the temperature distribution for a plane hot spot can be written

$$T(X, \varphi, t) = \sum_{m=1}^{\infty}\sum_{n=0}^{\infty} \exp(-\chi g_{mn}^2 t) \sin \lambda X (A_{mn} \cos n\varphi + B_{mn} \sin n\varphi), \tag{12}$$

where $\lambda = m\pi a/l$, $X = x/a$, $\chi = \lambda_0/c\varrho$, and m and n are the number of half waves in the x direction and the wave numbers in the φ direction, respectively,

$$g_{mn}^2 a^2 = \lambda^2 + n^2 + 2ka^2/\lambda_0 h$$

and A_{mn} and B_{mn} are found from the following expressions

$$A_{mn} = \frac{2a}{\pi l} \int_0^l \int_{-\pi}^{\pi} T(X, \varphi, 0) \sin \lambda X \cos n\varphi \, dX \, d\varphi , \qquad (n \neq 0) , \qquad A_{m0} = \frac{1}{2} (A_{mn})_{n=0} ,$$

$$B_{mn} = \frac{2a}{\pi l} \int_0^l \int_{-\pi}^{\pi} T(X, \varphi, 0) \sin \lambda X \sin n\varphi \, dX \, d\varphi , \qquad (n \neq 0) , \qquad B_{m0} = \frac{1}{2} (B_{mn})_{n=0} = 0 .$$

Considering equation (10) and using the property of delta function, we have

$$A_{mn} = \frac{2Q}{\pi la} \sin \lambda X_0 \cos n\varphi_0 , \qquad A_{m0} = \frac{Q}{\pi la} \sin \lambda X_0 ,$$

$$B_{mn} = \frac{2Q}{\pi la} \sin \lambda X_0 \sin n\varphi_0 .$$

Therefore

$$T(X, \varphi, t) = \frac{Q}{\pi la} \sum_{m=1}^{\infty} \sin \lambda X_0 \sin \lambda X \left[\exp(-\chi g_{m0}^2 t) + 2 \sum_{n=1}^{\infty} \exp(-\chi g_{mn}^2 t) \cos n(\varphi - \varphi_0) \right] . \tag{13}$$

In the same procedure we obtain a solution for a bending hot spot as

$$\overline{T}(X, \varphi, t) = \frac{\overline{Q}}{\pi lah} \sum_{m=1}^{\infty} \sin \lambda X_0 \sin \lambda X \left[\exp(-\chi \overline{g}_{m0}^2 t) + 2 \sum_{n=1}^{\infty} \exp(-\chi \overline{g}_{mn}^2 t) \cos n(\varphi - \varphi_0) \right] , \tag{14}$$

where

$$\overline{g}_{mn}^2 a^2 = \lambda^2 + n^2 + 6(2\lambda_0 + kh) a^2/\lambda_0 h^2 .$$

Here let us consider the convergence of the series in (13) or (14). The series of the type

$$\theta_3(y, \tau) = 1 + 2 \sum_{n=1}^{\infty} \exp(-\pi n^2 \tau) \cos 2n\pi y$$

is a theta function [6], which converges rapidly for large value of τ. On the other hand, for small value of τ the convergence of the series is rather poor. To improve this, JACOBI's imaginary transformation is applied [14], and then we have

$$\theta_3(y, \tau) = \sqrt{\frac{1}{\tau}} \exp\left(-\frac{\pi y^2}{\tau}\right) \left[1 + 2 \sum_{n=1}^{\infty} \exp\left(-\frac{\pi n^2}{\tau}\right) \cosh\left(\frac{2\pi n y}{\tau}\right) \right] .$$

It is clear that this series converges very rapidly for small τ. Another series in (13) or (14) with the form of

$$\sum_{m=1}^{\infty} \exp(-\chi \lambda^2 t) \sin \lambda X_0 \sin \lambda X$$

can also be replaced for small t by the expression as

$$\frac{1}{4} \sqrt{\frac{1}{\pi t'}} \left[\exp\left[-\frac{1}{4t'} \left(\frac{a}{l}\right) (X_0 - X)^2 \right] \left\{ 1 + 2 \sum_{m=1}^{\infty} \exp\left(-\frac{m^2}{t'}\right) \cosh\left[\frac{m}{t'} \left(\frac{a}{l}\right) (X_0 - X) \right] \right\} - \right.$$
$$\left. - \exp\left[-\frac{1}{4t'} \left(\frac{a}{l}\right) (X_0 + X)^2 \right] \left\{ 1 + 2 \sum_{m=1}^{\infty} \exp\left(-\frac{m^2}{t'}\right) \cosh\left[\frac{m}{t'} \left(\frac{a}{l}\right) (X_0 + X) \right] \right\} \right] , \qquad t' = \frac{\chi t}{(l/a)^2} .$$

Next let us consider a heat quantity stored in a cylindrical shell at time t. Denote the heat quantities by M for a plane hot spot and by $\overline{M}$ for a bending one, then we have [15]

$$M = \iiint_V c\varrho T_e(x, \varphi, z, t) \, dV = c\varrho ah \int_0^l \int_{-\pi}^{\pi} T(x, \varphi, t) \, dx \, d\varphi , \tag{15}$$

$$\overline{M}h = \iiint_V c\varrho z T_e(x, \varphi, z, t) \, dV = \frac{c\varrho ah^3}{12} \int_0^l \int_{-\pi}^{\pi} \overline{T}(x, \varphi, t) \, dx \, d\varphi , \qquad V\text{: Volume} . \tag{16}$$

Applying (13) to (15) and (14) to (16), we get

$$M = \frac{4c\varrho hQ}{\pi} \exp(-2\chi kt/\lambda_0 h) \sum_{m=1,3,5,\ldots}^{\infty} \frac{\exp(-\chi \lambda^2 t/a^2)}{m} \sin \lambda X_0 , \tag{17}$$

$$\overline{M} = \frac{c\varrho h\overline{Q}}{3\pi} \exp[-6\chi(2\lambda_0 + kh) t/\lambda_0 h^2] \sum_{m=1,3,5,\ldots}^{\infty} \frac{\exp(-\chi \lambda^2 t/a^2)}{m} \sin \lambda X_0 . \tag{18}$$

Denoting the heat quantities at $t = 0$ by M_0 and $\bar{M}_0$, respectively, and using the formula

$$\sum_{m=1,3,5,\dots}^{\infty} \frac{\sin \lambda X_0'}{m} = \frac{\pi}{4},$$

we find the relations between the intensity of the hot spot and the heat quantity as follows:

$$Q = M_0/c\varrho h\,, \qquad \bar{Q} = 12\bar{M}_0/c\varrho h\,. \tag{19), (20}$$

3.2 Solutions of equations of motion

Considering the boundary conditions (9a) and (9b), and the temperature distributions (13) and (14), the components of displacement may be expressed in the following form:

$$u(X, \varphi, t; x_0, \varphi_0) = \sum_{m=1}^{\infty} \sum_{n=0}^{\infty} U_{mn}(t) \sin \lambda X_0 \cos \lambda X \cos n(\varphi - \varphi_0)\,, \tag{21a}$$

$$v(X, \varphi, t; x_0, \varphi_0) = \sum_{m=1}^{\infty} \sum_{n=1}^{\infty} V_{mn}(t) \sin \lambda X_0 \sin \lambda X \sin n(\varphi - \varphi_0)\,, \tag{21b}$$

$$w(X, \varphi, t; x_0, \varphi_0) = \sum_{m=1}^{\infty} \sum_{n=0}^{\infty} W_{mn}(t) \sin \lambda X_0 \sin \lambda X \cos n(\varphi - \varphi_0)\,, \tag{21c}$$

where $U_{mn}(t)$, $V_{mn}(t)$ and $W_{mn}(t)$ are functions of time t. The initial conditions which should be specified are

$$u(X, \varphi, 0) = v(X, \varphi, 0) = w(X, \varphi, 0) = 0\,, \tag{22a}$$

$$\left[\frac{\partial}{\partial t} u(X, \varphi, t)\right]_{t=0} = \left[\frac{\partial}{\partial t} v(X, \varphi, t)\right]_{t=0} = \left[\frac{\partial}{\partial t} w(X, \varphi, t)\right]_{t=0} = 0\,. \tag{22b}$$

Let us examine the solutions for nonaxisymmetric mode ($n \neq 0$), since those for axisymmetric mode ($n = 0$) can be obtained more easily.

Applying the FOURIER transform to equations (6a) through (6c) under the consideration of (9a), (9b), (13), (14), (21a) (21b) and (21c), the displacements in terms of time t may be given as follows:

$$\gamma^2 \frac{d^2}{dt^2} \begin{bmatrix} U_{mn}(t) \\ V_{mn}(t) \\ W_{mn}(t) \end{bmatrix} + \begin{bmatrix} a_{11} & a_{12} & a_{13} \\ a_{12} & a_{22} & a_{23} \\ a_{13} & a_{23} & a_{33} \end{bmatrix} \begin{bmatrix} U_{mn}(t) \\ V_{mn}(t) \\ W_{mn}(t) \end{bmatrix} = \frac{2(1+\nu)\,\alpha Q}{\pi l} \exp(-\chi g_{mn}^2 t) \begin{bmatrix} \lambda \\ -n \\ -1 - \varkappa\lambda^2 \end{bmatrix} +$$
$$+ \frac{(1+\nu)\,\alpha h \bar{Q}}{6\pi l a} \exp(-\chi \bar{g}_{mn}^2 t) \begin{bmatrix} -\lambda \\ n \\ \lambda^2 + n^2 \end{bmatrix}, \tag{23}$$

where

$$a_{11} = -\lambda^2 - \frac{1-\nu}{2} n^2 - \frac{1-\nu}{2} \varkappa n^2\,, \qquad a_{12} = \frac{1+\nu}{2} \lambda n\,, \qquad a_{13} = \nu\lambda + \varkappa\left(\lambda^3 - \frac{1-\nu}{2} \lambda n^2\right),$$

$$a_{22} = -\frac{1-\nu}{2} \lambda^2 - n^2 - \frac{3(1-\nu)}{2} \varkappa\lambda^2\,, \qquad a_{23} = -n - \frac{3-\nu}{2} \varkappa \lambda^2 n\,,$$

$$a_{33} = -1 - \varkappa(\lambda^2 + n^2)^2 + (2n^2 - 1)\,\varkappa\,.$$

Applying LAPLACE transform to (23) under the initial conditions (22a) and (22b), and then solving the resulting equations, we obtain

$$\begin{bmatrix} U_{mn}^*(p) \\ V_{mn}^*(p) \\ W_{mn}^*(p) \end{bmatrix} \equiv \int_0^t \exp(-pt) \begin{bmatrix} U_{mn}(t) \\ V_{mn}(t) \\ W_{mn}(t) \end{bmatrix} dt = \frac{2(1+\nu)\,\alpha Q}{\pi l} \frac{1}{(p + \chi g_{mn}^2)\,\Delta} \times$$
$$\times \begin{bmatrix} u_1 & u_2 & u_3 \\ v_1 & v_2 & v_3 \\ w_1 & w_2 & w_3 \end{bmatrix} \begin{bmatrix} \gamma^4 p^4 \\ \gamma^2 p^2 \\ 1 \end{bmatrix} + \frac{(1+\nu)\,\alpha h \bar{Q}}{6\pi l a} \frac{1}{(p + \chi \bar{g}_{mn}^2)\,\Delta} \begin{bmatrix} \bar{u}_1 & \bar{u}_2 & \bar{u}_3 \\ \bar{v}_1 & \bar{v}_2 & \bar{v}_3 \\ \bar{w}_1 & \bar{w}_2 & \bar{w}_3 \end{bmatrix} \begin{bmatrix} \gamma^4 p^4 \\ \gamma^2 p^2 \\ 1 \end{bmatrix}, \tag{24}$$

where

$$\Delta = -p^6\gamma^6 + p^4\gamma^4 \cdot \text{trace}\, a_{ij} - p^2\gamma^2 \cdot \tfrac{1}{2}\,(a_{ii}a_{jj} - a_{ij}a_{ji}) + \det a_{ij}\,, \tag{25}$$

which can also be written as

$$\Delta = -(p^2\gamma^2 + \Omega_{mn1}^2)\,(p^2\gamma^2 + \Omega_{mn2}^2)\,(p^2\gamma^2 + \Omega_{mn3}^2)\,. \tag{26}$$

$\Delta = 0$, a characteristic equation, gives the dimensionless frequencies Ω_{mn1}^2, Ω_{mn2}^2 and Ω_{mn3}^2. $u_1, u_2, \dots, w_2, w_3$ and $\bar{u}_1, \bar{u}_2, \dots, \bar{w}_2, \bar{w}_3$, trace $a_{ij}, \dots,$ det a_{ij} can be calculated directly. Taking w_3, $\bar{w}_3$ and det a_{ij} for example, their concrete forms are given as

$$1 \gg \varkappa$$

$$\det a_{ij} = -\frac{1-\nu}{2}\varkappa\left[(\lambda^2+n^2)^4 - 2\nu\lambda^6 - 6\lambda^4 n^2 - 2(4-\nu)\lambda^2 n^4 - 2n^6 + \frac{1-\nu^2}{\varkappa}\lambda^4 + \right.$$

$$\left. + 2(2-\nu)\lambda^2 n^2 + n^4\right], \tag{27}$$

$$w_3 = \frac{(1-\nu)^2}{2}\left[-\frac{7}{2}\varkappa^2\lambda^4 n^2 + \frac{1}{2}\varkappa^2\lambda^2 n^4 - \lambda^2(\lambda^2+n^2)\right], \tag{28}$$

$$\overline{w}_3 = \frac{1-\nu}{2}\left[(\lambda^2+n^2)^3 - \nu\lambda^4 - (2+\nu)\lambda^2 n^2 + \lambda^2 n^2 - n^4\right]. \tag{29}$$

The inverse LAPLACE transform of (24) gives

$$\begin{bmatrix} U_{mn}(t) \\ V_{mn}(t) \\ W_{mn}(t) \end{bmatrix} = \sum_{i=1}^{3} \frac{1}{[\Omega_{mni}^2 + (\chi g_{mn}^2 \gamma)^2]\,[3\Omega_{mni}^4 + 2\,\mathrm{trace}\,a_{kj}\cdot\Omega_{mni}^2 + (a_{kk}a_{jj} - a_{jk}a_{jk})/2]} \times$$

$$\times \left\{ \frac{2(1+\nu)\alpha Q}{\pi l} \begin{bmatrix} u_1 & u_2 & u_3 \\ v_1 & v_2 & v_3 \\ w_1 & w_2 & w_3 \end{bmatrix} \begin{bmatrix} \Omega_{mni}^4 \\ -\Omega_{mni}^2 \\ 1 \end{bmatrix} \left[\exp(-\chi g_{mn}^2 t) + \frac{\chi g_{mn}^2 t}{\Omega_{mni}} \sin\frac{\Omega_{mni}}{\gamma} t - \cos\frac{\Omega_{mni}}{\gamma} t\right] + \right.$$

$$\left. + \frac{(1+\nu)\alpha h\overline{Q}}{6\pi l a} \begin{bmatrix} \overline{u}_1 & \overline{u}_2 & \overline{u}_3 \\ \overline{v}_1 & \overline{v}_2 & \overline{v}_3 \\ \overline{w}_1 & \overline{w}_2 & \overline{w}_3 \end{bmatrix} \begin{bmatrix} \Omega_{mni}^4 \\ -\Omega_{mni}^2 \\ 1 \end{bmatrix} \left[\exp(-\chi \overline{g}_{mn}^2 t) + \frac{\chi \overline{g}_{mn}^2 t}{\Omega_{mni}} \sin\frac{\Omega_{mni}}{\gamma} t - \cos\frac{\Omega_{mni}}{\gamma} t\right] \right\}. \tag{30}$$

The stress resultants and the stress couples can be found by substituting (21a) through (21c), together with (30), and (13) and (14) into (4a) through (4h). In general the solutions are obtained by imposing the weight for the results of a plane hot spot and a bending one.

Let us examine some simplified solutions of this kind of problem. The neglect of $\partial^2 u/\partial t^2$ and $\partial^2 v/\partial t^2$ in (6a) and (6b), which has often been made in the forced vibration analysis of shell problems, leads to the solutions of neglecting the in-plane inertia. In this case more simplified solutions are obtained. That is to say, the frequencies are given by

$$\Omega_{mn}^2 = \det a_{ij}/\det\,(\text{cofactor of } a_{33}),$$

and $u_1, v_1, w_1, w_2, \overline{u}_1, \overline{v}_1, \overline{w}_1, \overline{w}_2$ are equal to zero, and $u_2, v_2, \overline{u}_2, \overline{v}_2$ becomes more simple, while $u_3, v_3, w_3, \overline{u}_3, \overline{v}_3, \overline{w}_3$ are unchanged.

The elimination of all inertia terms in (6a) through (6c) enables us to get the solutions for the so-called quasi-stationary problem, which are given as

$$\begin{bmatrix} U_{mn}(t) \\ V_{mn}(t) \\ W_{mn}(t) \end{bmatrix} = \frac{1}{\det a_{ij}}\left[\frac{2(1+\nu)\alpha Q}{\pi l}\exp(-\chi g_{mn}^2 t)\begin{bmatrix} u_3 \\ v_3 \\ w_3 \end{bmatrix} + \frac{(1+\nu)\alpha h\overline{Q}}{6\pi l a}\exp(-\chi \overline{g}_{mn}^2 t)\begin{bmatrix} \overline{u}_3 \\ \overline{v}_3 \\ \overline{w}_3 \end{bmatrix}\right]. \tag{31}$$

The quasistatic solutions can be obtained by putting $t = 0$ in the above equations.

3.3 Thermal singularities

Theoretical interest provokes the examination of the thermal singularities caused by a hot spot. Let us consider the property of the solutions for a quasistatic problem. Take the radial displacements as an example, then by using the relations (27) through (29) we have for nonaxisymmetric modes

$$\begin{bmatrix} w(X, \varphi; X_0, \varphi_0) \\ \overline{w}(X, \varphi; X_0, \varphi_0) \end{bmatrix} =$$

$$= \sum_{m=1}^{\infty}\sum_{n=1}^{\infty} \frac{1}{(\lambda^2+n^2)^4 - 2\nu\lambda^6 - 6\lambda^4 n^2 - 2(4-\nu)\lambda^2 n^4 - 2n^6 + \frac{1-\nu^2}{\varkappa}\lambda^4 + 2(2-\nu)\lambda^2 n^2 + n^4} \times$$

$$\times \left[\frac{(1-\nu^2)\alpha Q}{\pi l \varkappa}\left[7\varkappa^2\lambda^4 n^2 - \varkappa^2\lambda^2 n^4 + 2\lambda^2(\lambda^2+n^2)\right] - \right.$$

$$\left. - \frac{2(1+\nu)\alpha a\overline{Q}}{\pi l h}\left[(\lambda^2+n^2)^3 - \nu\lambda^4 - (2+\nu)\lambda^2 n^2 + \lambda^2 n^2 - n^4\right]\right] \sin\lambda X_0 \sin\lambda X \cos n(\varphi - \varphi_0). \tag{32}$$

The denominator of (32) may be written

$$(\lambda^2 + n^2)^4 + \Psi(\text{terms of } \lambda^6, \lambda^4 n^2, \dots, n^4) .$$

Using this, the following relation holds

$$\frac{1}{(\lambda^2+n^2)^4+\Psi} = \frac{1}{(\lambda^2+n^2)^4} - \frac{\Psi}{(\lambda^2+n^2)^4\,[(\lambda^2+n^2)^4+\Psi]},$$

which suggests that the higher order terms with respect to the power of λ and/or n are represented in the first term of the right-hand side in the above expression.

The higher order terms of (32) are:

for plane hot spot;

$$\sum_{m=1}^{\infty}\sum_{n=1}^{\infty} \frac{7\lambda^4 n^2 - \lambda^2 n^4}{(\lambda^2+n^2)^4} \sin\lambda X_0 \sin\lambda X \cos n(\varphi - \varphi_0) \quad \text{for FLÜGGE}, \tag{33a}$$

$$\sum_{m=1}^{\infty}\sum_{n=1}^{\infty} \frac{\lambda^2}{(\lambda^2+n^2)^3} \sin\lambda X_0 \sin\lambda X \cos n(\varphi - \varphi_0) \quad \text{for DONNELL}, \tag{33b}$$

for bending hot spot;

$$\sum_{m=1}^{\infty}\sum_{n=1}^{\infty} \frac{1}{\lambda^2+n^2} \sin\lambda X_0 \sin\lambda X \cos n(\varphi - \varphi_0) \quad \text{for FLÜGGE and DONNELL}. \tag{34}$$

For simplicity let us take (34) as an example:

$$\sum_{n=1}^{\infty} \frac{\cos n(\varphi-\varphi_0)}{n^2+\lambda^2} = -\frac{1}{2\lambda^2} + \frac{\pi}{2\lambda}\cdot\frac{\cosh\lambda(\varphi-\varphi_0-\pi)}{\sinh\lambda\pi}, \qquad 0 \leqq \varphi - \varphi_0 \leqq 2\pi . \tag{35}$$

Using this relation, (34) is rewritten for nonaxisymmetric term as

$$\frac{\pi}{2}\sum_{m=1}^{\infty} \frac{\cosh\lambda(\varphi-\varphi_0-\pi)}{\lambda\sinh\lambda\pi} \sin\lambda X_0 \sin\lambda X .$$

This expression is again transformed as

$$\frac{\pi}{2}\sum_{m=1}^{\infty} \frac{\exp(-\lambda\pi)\cosh\lambda(\varphi-\varphi_0)}{\lambda\sinh\lambda\pi} \sin\lambda X_0 \sin\lambda X + \frac{\pi}{2}\sum_{m=1}^{\infty}\frac{\exp[-\lambda(\varphi-\varphi_0)]}{\lambda} \sin\lambda X_0 \sin\lambda X . \tag{36}$$

The first series is uniformly convergent, and the second may be transformed if we employ the FOURIER expansion of the function:

$$\sum_{m=1}^{\infty} \frac{\exp[-\lambda(\varphi-\varphi_0)]}{\lambda} \sin\lambda X_0 \sin\lambda X = -\frac{l}{4\pi a}\left\{\ln 2\left[\cosh\frac{\pi a}{l}(\varphi-\varphi_0) - \cos\frac{\pi a}{l}(X-X_0)\right] - \right.$$

$$\left. - \ln 2\left[\cosh\frac{\pi a}{l}(\varphi-\varphi_0) - \cos\frac{\pi a}{l}(X+X_0)\right]\right\}. \tag{37}$$

It can be noticed at once that the first term in the right-hand side of this expression tends to infinity as $X \to X_0$ and $\varphi \to \varphi_0$. In more detail, the power expansion of this term gives

$$-\frac{l}{4\pi a}\ln\left[\left(\frac{\pi a}{l}\right)^2(\varphi-\varphi_0)^2 + \left(\frac{\pi a}{l}\right)^2(X-X_0)^2 + \dots\right]$$

and hence we understand

$$-\frac{l}{2\pi a}\ln\frac{\pi a}{l}(\varphi-\varphi_0) \quad \text{at} \quad X = X_0 ,$$

$$-\frac{l}{2\pi a}\ln\frac{\pi a}{l}(X-X_0) \quad \text{at} \quad \varphi = \varphi_0 .$$

They correspond to the order of singularity. That is to say, it follows that the radial displacement for a bending hot spot has a singularity of

$$\frac{(1+\nu)\,\alpha\bar{Q}}{2\pi h}\ln\frac{\pi a}{l}(\varphi-\varphi_0) \quad \text{and/or} \quad \frac{(1+\nu)\,\alpha\bar{Q}}{2\pi h}\ln\frac{\pi a}{l}(X-X_0) . \tag{38}$$

Performing the same procedure for (32a) and (33b), we recognize that (33a) also exhibits the same order of singularity as a bending hot spot, while (33b), the radial displacement based on DONNELL theory for a plane hot spot, does no t reveal any singularity property.

The results of higher order of singularities at $\varphi = \varphi_0$ of the displacements, stress resultants and the stress couples are listed in Table 1 for a plane hot spot and in Table 2 for a bending one, respectively.

As for the property of the singularities of the membrane stresses and the bending stresses, we can find some interesting characters.

Table 1. Thermal singularities at $\varphi = \varphi_0$ for a plane hot spot

	FLÜGGE	DONNELL
u	$\frac{(1+\nu)\,\alpha Q}{2\pi a(X - X_0)}$	$\frac{(1+\nu)\,\alpha Q}{2\pi a(X - X_0)}$
v	0	0
w	$-\frac{(1-\nu^2)\,\alpha Q}{128\pi a}\left(\frac{h}{a}\right)^2 \ln\frac{\pi a}{l}\,\lvert X - X_0\rvert$	—
N_x	$-\frac{E\alpha h Q}{2\pi a^2(X - X_0)^2}$	$-\frac{E\alpha h Q}{2\pi a^2(X - X_0)^2}$
N_φ	$\frac{E\alpha h Q}{2\pi a^2(X - X_0)^2}$	$\frac{E\alpha h Q}{2\pi a^2(X - X_0)^2}$
M_x	$\frac{E\alpha Q}{24\pi(X - X_0)^2}\left(\frac{h}{a}\right)^3$	$\frac{(3+\nu)\,E\alpha Q}{16\pi}\left(\frac{h}{a}\right)\ln\frac{\pi a}{l}\,\lvert X - X_0\rvert$
M_φ	$\frac{(1+3\nu)\,E\alpha Q}{4608\pi(X - X_0)^2}\left(\frac{h}{a}\right)^5$	$\frac{(1+3\nu)\,E\alpha Q}{16\pi}\left(\frac{h}{a}\right)\ln\frac{\pi a}{l}\,\lvert X - X_0\rvert$

Table 2. Thermal singularities at $\varphi = \varphi_0$ for a bending hot spot

	FLÜGGE	DONNELL
$\bar{u}$	$-\frac{(1+\nu)\,\alpha\bar{Q}}{24\pi a(X - X_0)}\left(\frac{h}{a}\right)$	$-\frac{(1+\nu)^2\,\alpha\bar{Q}}{48(1-\nu)\,\pi a(X - X_0)}\left(\frac{h}{a}\right)$
$\bar{v}$	0	0
$\bar{w}$	$\frac{(1+\nu)\,\alpha\bar{Q}}{2\pi h}\ln\frac{\pi a}{l}\,\lvert X - X_0\rvert$	$\frac{(1+\nu)\,\alpha\bar{Q}}{2\pi h}\ln\frac{\pi a}{l}\,\lvert X - X_0\rvert$
$\bar{N}_x$	$-\frac{E\alpha\bar{Q}}{12\pi a(X - X_0)^2}\left(\frac{h}{a}\right)^2$	$-\frac{(1+3\nu)\,E\alpha\bar{Q}}{4(1-\nu)\,\pi a(X - X_0)^2}\left(\frac{h}{a}\right)^2$
$\bar{N}_\varphi$	$\frac{E\alpha\bar{Q}}{24\pi a(X - X_0)^2}\left(\frac{h}{a}\right)^2$	$-\frac{(1+\nu)\,E\alpha\bar{Q}}{48(1-\nu)\,\pi a(X - X_0)^2}\left(\frac{h}{a}\right)^2$
$\bar{M}_x$	$-\frac{E\alpha\bar{Q}}{24\pi(X - X_0)^2}\left(\frac{h}{a}\right)^2$	$-\frac{E\alpha\bar{Q}}{24\pi(X - X_0)^2}\left(\frac{h}{a}\right)^2$
$\bar{M}_\varphi$	$\frac{E\alpha\bar{Q}}{24\pi(X - X_0)^2}\left(\frac{h}{a}\right)^2$	$\frac{E\alpha\bar{Q}}{24\pi(X - X_0)^2}\left(\frac{h}{a}\right)^2$

According to FLÜGGE's theory we obtain for a plane hot spot

$$\frac{6M_x/h}{N_x} \to -\frac{1}{2}\left(\frac{h}{a}\right), \qquad \frac{6M_\varphi/h}{N_\varphi} \to \frac{1+3\nu}{384}\left(\frac{h}{a}\right).$$

This shows that the bending stresses have an order of singularity of (h/a) or $(h/a)^3$ compared with the membrane stresses. On the other hand we obtain for a bending hot spot

$$\frac{\bar{N}_x}{6\bar{M}_x/h} \to \frac{1}{3}\left(\frac{h}{a}\right), \qquad \frac{\bar{N}_\varphi}{6\bar{M}_\varphi/h} \to \frac{1}{6}\left(\frac{h}{a}\right)^3.$$

In this case the membrane stresses also have an order of singularity of (h/a) compared with the bending ones.

According to DONNELL's results, based on the assumption of shallow theory, the order of singularity of the bending stresses is lower than that of the membrane one for a plane hot spot, and for a bending hot spot we have

$$\frac{\bar{N}_x}{6\bar{M}_x/h} \to \frac{1+3\nu}{1-\nu}\left(\frac{h}{a}\right), \qquad \frac{\bar{N}_\varphi}{6\bar{M}_\varphi/h} \to -\frac{1+\nu}{12(1-\nu)}\left(\frac{h}{a}\right).$$

4. Numerical examples and discussion

As has been described in 3.3, a cylindrical shell under a hot spot exhibits the higher order of thermal singularities not only for the stresses but also even for the displacements, except for the radial displacement based on DONNELL theory under a plane hot spot. Therefore the practical application requires the results of a cylindrical shell under a line or an area heat source. The solutions for such a case can readily be obtained by the direct integration of the solutions obtained already, since they may be considered as the GREEN's function of the nonstationary problem of a cylindrical shell. That is to say, if $\Phi(x, \varphi, t; x_0, \varphi_0)$, the representative form of the displacements and the stresses, represents the quantity due to a hot spot with unit intensity acting at (x_0, φ_0), each quantity for the case where the heated portion is local, as shown in Figure 2, may be given as

$$\int_{x_0-\delta}^{x_0+\delta} \int_{\varphi_0-\varepsilon}^{\varphi_0+\varepsilon} q(x_0, \varphi_0)\, \Phi(x, \varphi, t; x_0, \varphi_0)\, dx_0\, a\, d\varphi_0\,, \tag{39}$$

where $q(x_0, \varphi_0)$ is the intensity of the distributed heat source. In the numerical calculations q is assumed to be uniform and again replaced with the total intensity Q $(= 4qa\delta\varepsilon)$.

Numerical examples are presented in order to examine the effects of the plane and/or the bending heat source, of the theories due to FLÜGGE and DONNELL, and of the dynamic, thermal effects on the response of a cylindrical shell. The center of the heated area is at the center of the cylindrical shell ($x_0/l = 1/2$, $\varphi_0 = 0$). The dimensions of the heated area are $\varepsilon a = (\pi/30)\, a$ in the φ direction and $\delta = \varepsilon a$ in the axial direction. POISSON's ratio ν is chosen to be 0.3.

The following dimensionless parameters are introduced to specify the response due to the time and dynamic, thermal effects.

$$\tau = t/\beta\,, \qquad B_i = kh/\lambda_0\,, \qquad \chi_\beta = \chi\beta/a^2\,, \qquad \beta = \sqrt{\varrho a^2 h^3/K}\,. \tag{40}$$

Then

$$\text{plane heat source:} \qquad \chi g^2_{mnl} t = \chi_\beta[\lambda^2 + n^2 + 2B_i(a/h)^2]\,\tau\,; \tag{41a}$$

$$\text{bending heat source:} \qquad \chi \bar{g}^2_{mnl} t = \chi_\beta[\lambda^2 + n^2 + 6(2 + B_i)\,(a/h)^2]\,\tau\,. \tag{41b}$$

χ_β, including the coefficient of thermal conductivity, plate velocity, etc., may be so small for the real material that it is chosen to be 10^{-6}. B_i is the BIOT number.

The responses of the radial displacement and the stresses at the center of the heated area ($x/l = 1/2$, $\varphi = 0$) are shown in the numerical results and the number of the maximum term of the double series taken in the numerical calculations is 199 for m and 100 for n, since the convergence of the series is rather poor for the stresses because of the local property of the heated area.

The dynamic effects can be significantly evaluated based on the quasistatic solutions, which were outlined in the last part of 3.2. In order to examine the variation of the radial displacement of a cylindrical shell under a plane heat source and a bending one, the values for the quasistatic analysis are shown in Figure 3 (a) and (b), in which the variations due to the length-to-radius ratio (l/a) and the radius-to-thickness ratio (a/h) are described together with the results based on the theories of FLÜGGE and DONNELL. It is apparent that the effect of the temperature gradient is none the less important compared with that of the mean temperature rise for the small values of a/h and l/a, and the application range of DONNELL's equation is limited to the small value of l/a, as being suggested by various studies for another problems such as static, forced vibration problems, etc. That the minimum value of a/h is taken to be 10 is based on the suggestion that most thin shell theories have been assumed to be valid up to this point [16].

Figures 4, 5 and 6 are representative of the dynamic response of a cylindrical shell under a respective heat source. Figures 4 show the radial displacement response. From (a) it is seen that the radial displacement oscillates about the quasistatic value with the period of oscillation corresponding to the mode wave which is related to the frequency of a cylindrical shell. In the case of $a/h = 50$ and $l/a = 5$, for example, the lowest frequency occurs for the mode wave of $m = 1$ and $n = 3$. The period of oscillation for this mode is nearly 30.6. The peak of the response occurs at the time a little shorter than the half of this period. For a short cylinder such as $l/a = 5$ the response is also affected by the combination of another mode waves, which makes the period of oscillation somewhat shorter and the wave pattern more complicated. The response due to DONNELL's equation is smaller than that due to FLÜG-

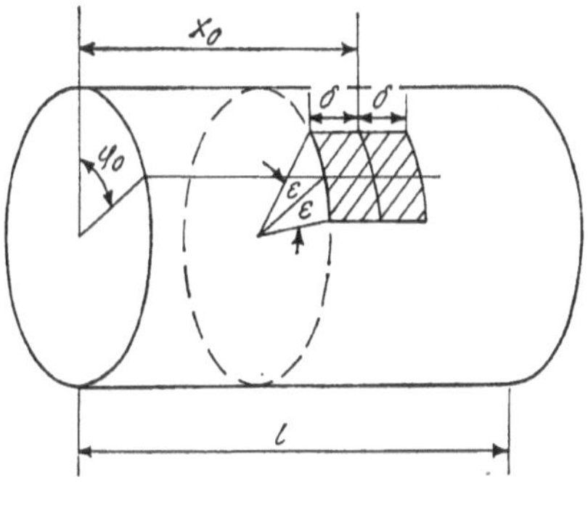

Fig. 2. Locally heated cylindrical shell

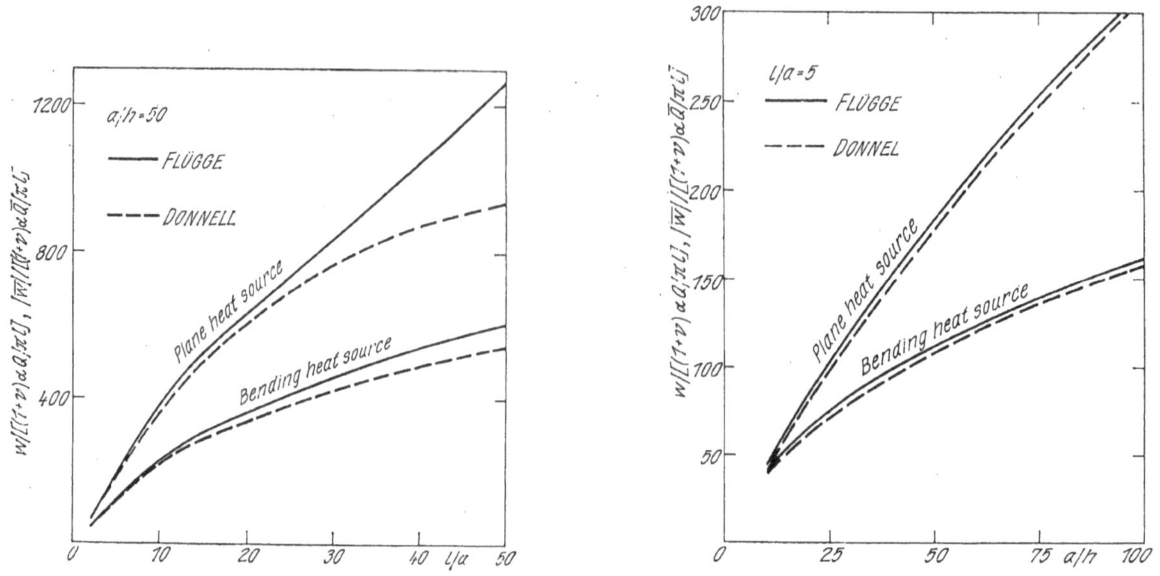

Fig. 3. Quasistatic radial displacement
(a) Variation due to l/a
(b) Variation due to a/h

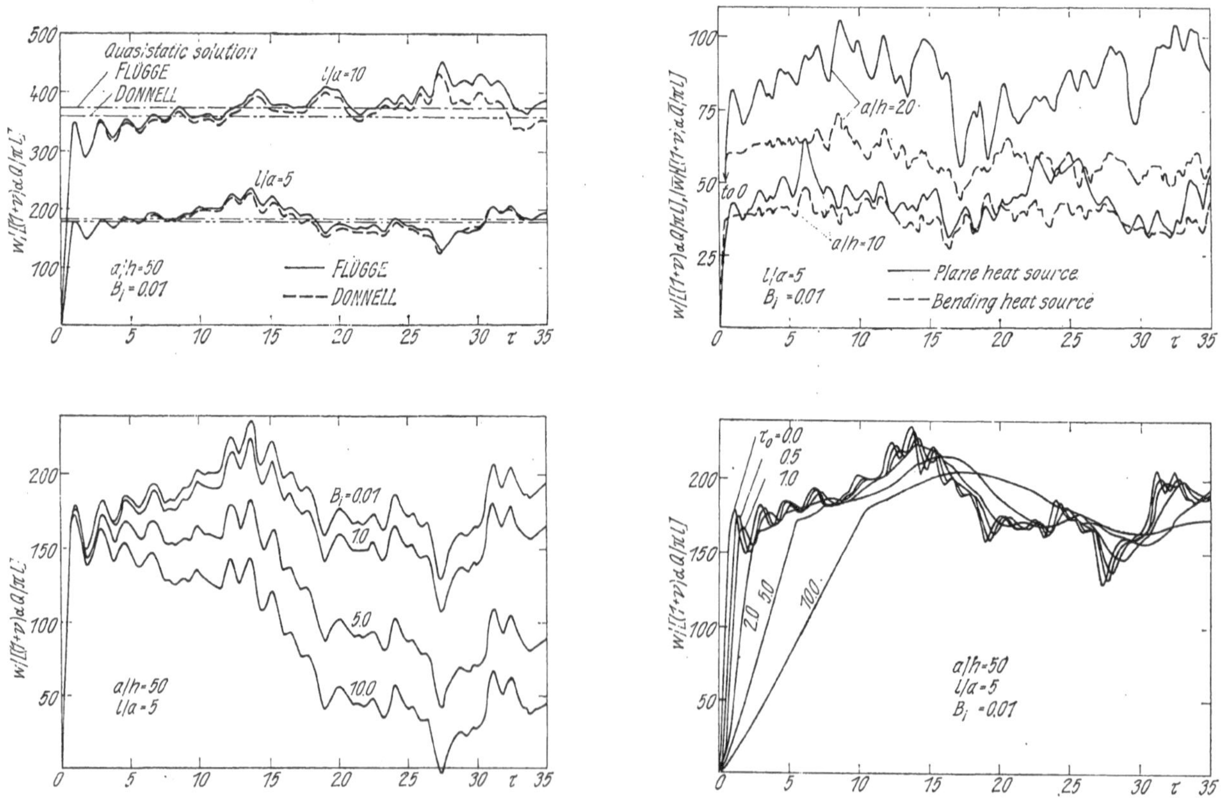

Fig. 4. Radial displacement response
(a) Response due to FLÜGGE and DONNELL theories
(b) Response to plane and bending heat sources (FLÜGGE)
(c) Response due to B_i (FLÜGGE)
(d) Response due to heated time τ_0 (FLÜGGE)

GE's one and the dynamic factor in both theories is approximately 1.2 to 1.3 for $l/a = 5$ and 10. The response due to the plane and bending heat sources are shown in Figure 4(b) together with the variation of a/h. The variation due to the BIOT number is shown in Figure 4(c). For small value of B_i the maximum displacement occurs nearly at the same time, while with the increase of B_i the displacement with the lapse of time becomes small, which suggests that in this case the response at an early stage should be considered.

The results obtained above are those for the most severe condition of an inertia effect. It may be general to consider the heat input to be applied gradually. If the heat quantity supplied at unit time is constant, the expressions taking into account the heated time t_0 may be obtained by the following replacement of the expressions found already:

temperature distribution for a plane heat source;

$$\exp(-\chi g_{mn}^2 t) \rightarrow \begin{cases} \dfrac{1 - \exp(-\chi g_{mn}^2 t)}{\chi g_{mn}^2 t_0}, & (0 \leqq t \leqq t_0), \quad (42\text{a}) \\ \dfrac{\exp(-\chi g_{mn}^2 t)\,[\exp(\chi g_{mn}^2 t_0) - 1]}{\chi g_{mn}^2 t_0}, & (t_0 \leqq t), \quad (42\text{b}) \end{cases}$$

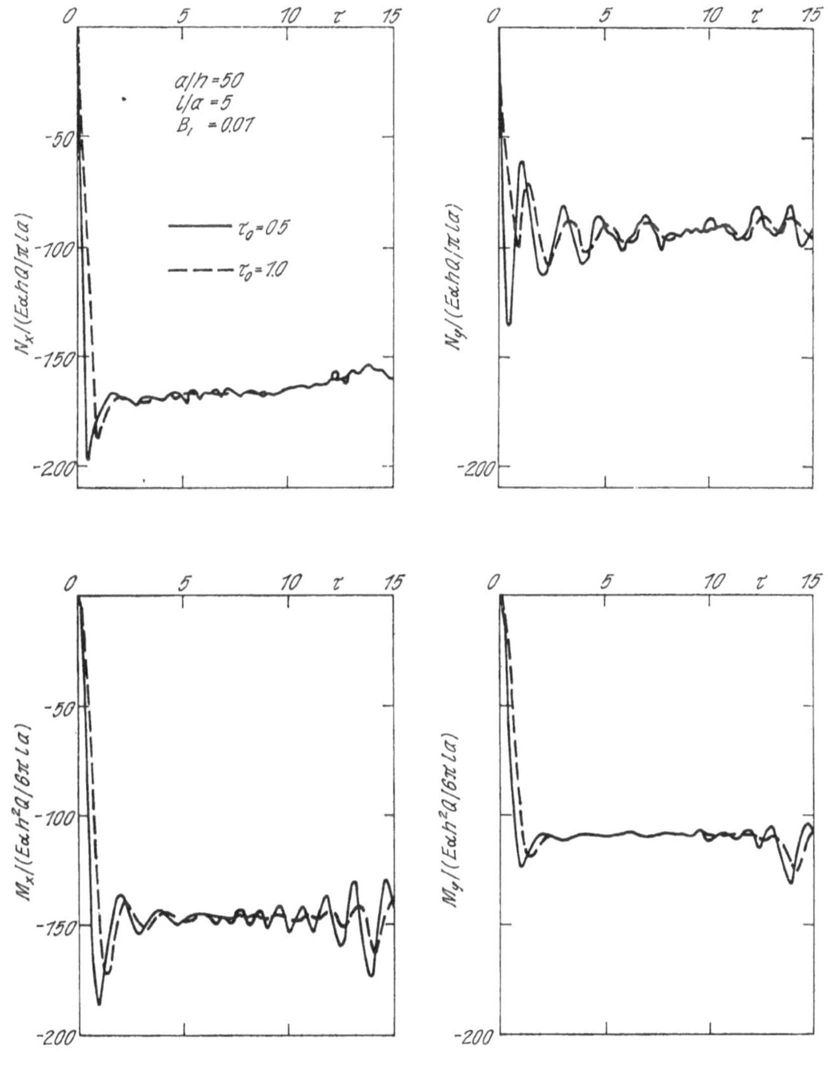

Fig. 5. Stress response to plane heat source
(a) axial membrane stress
(b) circumferential membrane stress
(c) axial bending stress
(d) circumferential bending stress

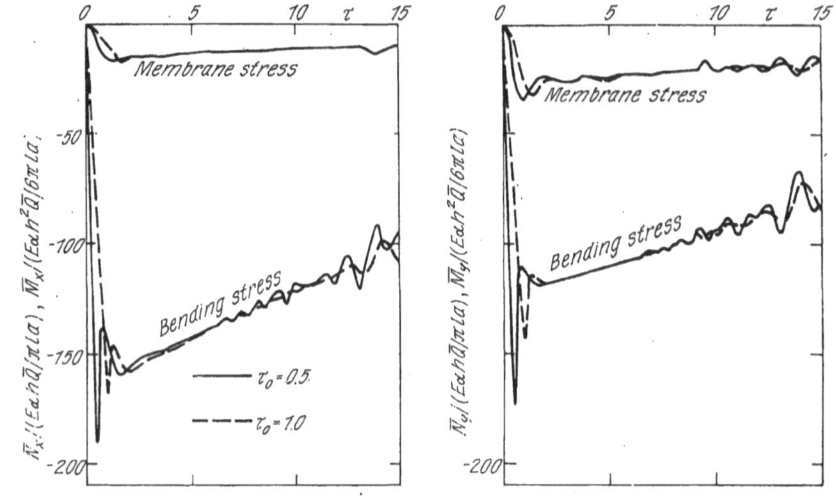

Fig. 6. Stress response to bending heat source (FLÜGGE)
(a) axial stress
(b) circumferential stress

displacements and stresses for a plane heat source;

$$\sum_{i=1}^{3} \frac{1}{(\Omega\gamma)^2 + (\chi g_{mn}^2\gamma)^2}\left[\exp\left(-\chi g_{mn}^2 t\right) + \frac{\chi g_{mn}^2}{\Omega}\sin\Omega t - \cos\Omega t\right] \rightarrow$$

$$\rightarrow \begin{cases} \dfrac{1}{\chi g_{mn}^2 t_0}\displaystyle\sum_{i=1}^{3}\left\{\frac{1}{\Omega^2\gamma^2}(1-\cos\Omega t) - \frac{1}{(\Omega\gamma)^2+(\chi g_{mn}^2\gamma)^2}\times\right. \\ \qquad \left.\times\left[\exp\left(-\chi g_{mn}^2 t\right) + \dfrac{\chi g_{mn}^2}{\Omega}\sin\Omega t - \cos\Omega t\right]\right\}, \qquad (0 \leqq t \leqq t_0)\,, & (43\text{a}) \\ \dfrac{1}{\chi g_{mn}^2 t_0}\displaystyle\sum_{i=1}^{3}\left[\frac{1}{\Omega^2\gamma^2}[\cos\Omega(t-t_0) - \cos\Omega t] - \frac{1}{(\Omega\gamma)^2+(\chi g_{mn}^2\gamma)^2}\left\{\exp\left(-\chi g_{mn}^2 t\right) - \exp\left[-\chi g_{mn}^2(t-t_0)\right] + \right.\right. \\ \left.\left. + \dfrac{\chi g_{mn}^2}{\Omega}[\sin\Omega t - \sin\Omega(t-t_0) - \cos\Omega t + \cos\Omega(t-t_0)]\right\}\right], \qquad (t_0 \leqq t)\,, \qquad \Omega = \Omega_{mni}/\gamma\,. & (43\text{b}) \end{cases}$$

The response of the radial displacement taking into account the heated time is shown in Figure 4 (d). It is apparent that the response for the small value of $\tau_0(=t_0/\beta)$ is severe.

Next let us consider the stress response. In order to examine the contribution of the membrane stress and the bending stress to the actual stress, each stress is separately given in the figures. The nondimensionalization of the membrane and bending stresses is made so as to give the inner surface stress by the direct addition and the outer surface stress by the direct subtraction. When τ_0 is equal to zero, only the terms including T or $\overline{T}$ in the expressions (4a) through (4f) appear, since all the displacements are zero at $\tau = 0$, that is to say, it follows that the stresses at the state of the severe condition of the inertia effect depend only on the temperature distribution at an initial state. This is not practical. Therefore the stresses should be discussed in consideration of the heated time. Figures 5 illustrate the response of the stresses based on the FLÜGGE's equation for a plane heat source. The results calculated by the DONNELL's equation are almost the same for the parameters selected here as those by FLÜGGE. N_x has its maximum compressive stress at $\tau = \tau_0$. On the other hand, as for N_φ in the case of $\tau_0 = 1.0$ the maximum compressive stress occurs at time other than $\tau = 1.0$, which suggests that dynamic effect plays also an important role in this kind of problem. The response of M_x and M_φ makes this fact more clear. The effect of T on M_x is so small that the dynamic effect becomes predominant and it causes comparatively large compressive stress in the range of $\tau = 13.5 \sim 14.0$ as in the case of the radial displacement response. As for M_φ so it is. It should be noted that in view of the order of the compressive stress, even for a plane heat source, M_x is almost of the same magnitude as N_x.

The stress response due to a bending heat source is shown in Figures 6, in which it is understood that even for a thin cylindrical shell $(a/h = 50)$ the maximum compressive stresses $\overline{M}_x$ and $\overline{M}_\varphi$, which are by far larger than the membrane stresses $\overline{N}_x$ and $\overline{N}_\varphi$, are of the same order of magnitude as the maximum compressive stresses due to a plane heat source.

In conclusion, it can be said that the dynamic treatment of a cylindrical shell under some heat applications is found to be important to understand the real behaviour of the structure, especially for rapidly applied heating.

References

1 FLÜGGE, W.; CONRAD, D. A., Thermal Singularities for Cylindrical Shells, Proc. 3rd U.S. Nat. Congr. Appl. Mech. pp. 321–328 (1958).
2 JAHANSHAHI, A., Thin Plates and Shallow Cylindrical Shells Subjected to Hot Spots, Trans. ASME, Ser. E, **31**, pp. 79–82 (1964).
3 WILSON, J. D., Thermal Stresses in Thin Cylindrical Shells, 2nd Int. Conf. on Struct. Mech. in Reactor Tech., Berlin, Gl/10 (1973).
4 SCHNELL, W.; WARDENBACH, W., Näherungsweise Ermittlung der Wärmespannungen in Kreiszylinderschalen, Ing.-Arch. **48**, pp. 417–430 (1979).
5 McQUILLEN E. J.; BRULL, M. A., Dynamic Thermoelastic Response of Cylindrical Shells, Trans. ASME, Ser. E, **37**, pp. 661–670 (1970).
6 HUANG, C. L. D.; WOO, H. K., The fundamental Heat Conduction Solution for a Thermal Shock on a Finite Orthotropic Cylindrical Shell, Nucl. Engng. and Design, **45**, pp. 67–79 (1978).
7 SHIRAKAWA, K., Thermoelastic Behavior of a Cylindrical Shell Subjected to an Instantaneous Hot Spot, Bull. Japan Soc. Mech. Engng., **17**, pp. 187–193 (1974).
8 SHIRAKAWA, K., Transient Response of Cylndrical Shells to Localized Heat Sources, Nucl. Engng. and Design, **54**, pp. 337–347 (1979).
9 FLÜGGE, W., Stresses in Shells, Springer-Verlag, 1960.
10 DONNELL, L. H., Stability of Thin-Walled Tubes under Torsion, Tech. Rep. No. 479, NACA, pp. 95–116 (1933).
11 BOLOTIN, V. V., Equations for the Non-Stationary Temperature Fields in Thin Shells in the Presence of Sources of Heat, PMM, **24**, pp. 515–519 (1960).
12 MIZOGUCHI, K., On Thermoelasticity of Cylindrical Shells, Bull. Univ. Osaka Pref. Ser. A, **15**, pp. 1–9 (1965).
13 TAKEUTI, Y., Thermal Stresses in Circular Disc due to Instantaneous Line Heat Source, ZAMM **45**, pp. 177–184 (1965).
14 Encyclopedia of Mathematics (in Japanese), 2nd Edition, Iwanami-Shoten, 1968.
15 MIZOGUCHI, K., Thermal Stresses in a Cylindrical Shell under a Moving Hot Spot, Bull. Japan Soc. Mech. Engng., **15**, pp. 184 to 192 (1972).
16 WEINGARTEN, L. I.; REISMANN, H., Forced Motion of Cylindrical Shells— A Comparison of Shell Theory with Elasticity Theory, ZAMM, **54**, pp. 181–191 (1974).

Eingereicht am 19. 6. 1980

Anschrift: Dr. KAORU SHIRAKAWA, Department of Mechanical Engineering, University of Osaka Prefecture, Mozu-Umemachi, Sakai, Osaka 591, Japan

ZAMM 61, 435–441 (1981)

S. K. Roy / J. P. Agarwal

Hydromagnetic Flow in Thin Films of Liquid with a Rheological Power-Law

Das Problem der Bestimmung von Strömungsparametern für den dünnen Film einer leitenden Ostwald-de Waele-Flüssigkeit, die an einer vertikalen Platte in Gegenwart eines magnetischen Feldes haftet, wurde für zwei Fälle untersucht: für die stationäre Platte und die mit konstanter Geschwindigkeit aus einem Flüssigkeitsbad herausgezogene Platte. Es zeigt sich, daß eine Verstärkung des Magnetfeldes bei der stationären Platte die Geschwindigkeit vermindert und bei der bewegten Platte die Geschwindigkeit erhöht. Der Durchsatz des Films sinkt jedoch mit Verstärkung des Magnetfeldes in beiden Fällen und die Filmströmung ist dabei eingeschränkt. Bei der Strömung eines dilatanten Flüssigkeitsfilms entlang der stationären Platte existiert ein kritischer Wert der verallgemeinerten Hartmann-Zahl, bei dessen Überschreitung in der Nähe der freien Oberfläche eine Zone auftritt, in der die Flüssigkeit wie ein quasifester Körper strömt. Diese Zone weitet sich mit Zunahme der verallgemeinerten Hartmann-Zahl aus.

The problem of determining the flow characteristics of a thin film of conducting non-Newtonian power-law liquid adhering to a vertical plate in the presence of a magnetic field has been investigated for two cases, namely, the plate stationary and the plate withdrawn with a constant velocity from a liquid bath. It is found that the effect of increase in magnetic field is to decrease the velocity in the case of stationary plate and to increase the velocity in the case of moving plate. The flow rate of the film, however, decreases with increase in the magnetic field in both the cases, and the flow of the film is thereby restricted. Moreover, when a dilatant liquid film flows along a stationary plate, there is found to exist a critical value of the generalized Hartmann number, and if the generalized Hartmann number exceeds its critical value, there appears a zone near the free surface in which the liquid flows like a quasi-solid body. This zone widens with the increase in the generalized Hartmann number.

Исследуется вопрос определения параметров течения для тонкого слоя проводящей неньютоновской жидкости Оствалдаде Вэле, прилипающийся к вертикальной пластинке в присутствии магнитного поля. Рассмотрено два случая: для стационарной пластинки и для пластинки, которую вытаскивают с постоянной скоростью из сосуда, наполненный жидкостью. Увеличение магнитного поля в случае стационарной пластинки приводит к уменьшению скорости, а в случае движущейся пластинки к росту скорости течения. В обоих случаях уменьшается расход жидкости тонкого слоя и течение подвергается определённым ограничениям. При течении дилатантного жидкостного слоя вдоль стационарной пластинки существует критическое значение обобщённого числа Хартмана, выше которого вблизи свободной поверхности образуется область, в которой жидкость движется как квазитвёрдое тело. Эта область расширяется с ростом обобщённого числа Хартмана.

1. Introduction

Several considerations like expansion of the applications of MHD, the possibility of using new working substances in various MHD devices, the presence of all the rheological properties in various degrees in every real material create a need for investigation of MHD flow in conducting media with more complex physical and mechanical properties than those of an incompressible viscous Newtonian fluid. Thus, there have recently been published a number of research papers which investigate MHD flows of conducting non-Newtonian fluids. These investigations have revealed a number of effects related to the presence of non-Newtonian properties in the fluid, for instance, in strong magnetic fields, even in the presence of insignificant deviations from Newtonian properties, the character of the flow of the fluid can differ considerably from the flow of a Newtonian fluid [1].

Up to this time a number of problems of MHD flows have been studied in several non-Newtonian fluids, in particular, the one with a rheological power-law which adequately approximates the properties of a number of fluids, and is widely used in the theoretical rheology. As it is, the Hartmann flow of a power-law fluid was investigated by Sarpkaya [2], and it was found that the velocity increases with increase in the magnetic field. However, later on, the analysis of Martinson and Pavlov [3] revealed the formation of quasi-solid zone in the flow of a dilatant fluid for values of the Hartmann number, generalized to the case of the media, exceeding some critical value, and this effect was named as magnetic plasticity effect. Furthermore, it was noted by Martinson and Pavlov [1] that in the case of conducting dilatant fluids, the shear disturbances propagating with finite velocity have to be localized in a finite region of space. These findings, thus, explained the appearance of quasi-solid zones in steady flows of conducting dilatant fluids mentioned earlier. Again, the investigation of the problem of hydrodynamic stability of plane flows of a conducting dilatant fluid in a transverse magnetic field by Pavlov [4] showed the formation of a finite zone in which tangential shear stresses are localized.

Here, we consider the problem of determining the flow characteristics of a thin film of conducting non-Newtonian liquid with a rheological power-law adhering to a vertical wall in the presence of a transverse magnetic field for the two cases, namely, the wall stationary and the wall moving with a constant velocity. It is found that the effect of increase in magnetic field is to decrease the velocity in the case of plate at rest and to increase the velocity in the case of moving plate. The flow rate of the film, however, decreases with increase in the magnetic field in both the cases, and the fall of the film is thus restricted. Additionally, as expected, in the case of a dilatant liquid film, there is found to exist a critical value of the generalized Hartmann number beyond which there appears a quasi-solid zone near the free surface.

2. Formulation of the problem

The steady motion of an incompressible, electrically conducting liquid with a rheological power-law in the presence of a magnetic field is governed by the following equations:

Maxwell's equations:

$$\operatorname{curl} \boldsymbol{H} = \boldsymbol{J}, \qquad \operatorname{curl} \boldsymbol{E} = 0, \qquad \operatorname{div} \boldsymbol{B} = 0; \tag{2.1), (2.2), (2.3}$$

Ohm's law:

$$\boldsymbol{J} = \sigma_e(\boldsymbol{E} + \boldsymbol{V} \times \boldsymbol{B}), \tag{2.4}$$

where

$$\boldsymbol{B} = \mu_m \boldsymbol{H}; \tag{2.5}$$

Equation of continuity:

$$\operatorname{div} \boldsymbol{V} = 0; \tag{2.6}$$

Momentum equation:

$$\varrho(\boldsymbol{V} \cdot \operatorname{grad} \boldsymbol{V}) = -\operatorname{grad} p + \nabla \cdot \tau + \boldsymbol{F} \tag{2.7}$$

with τ, the shear stress expressed in Cartesian coordinate system, given by

$$\tau = k \left|\frac{\mathrm{d}u}{\mathrm{d}y}\right|^{n'-1} \frac{\mathrm{d}u}{\mathrm{d}y} \qquad (n' > 0), \tag{2.8}$$

and $\boldsymbol{F}$, the external force.

Here the various symbols used have their usual meanings.

If we take the x-axis vertically downwards and parallel to the plate and the y-axis normal to the plate, and suppose that the velocity $\boldsymbol{V}$ is in x-direction, and a uniform magnetic field B_0 is imposed perpendicular to the plate, then the equation of motion for the flow in a thin film of liquid adhering to a vertical plane surface simplifies to

$$n' \left|\frac{\mathrm{d}u}{\mathrm{d}y}\right|^{n'-1} \frac{\mathrm{d}^2u}{\mathrm{d}y^2} - \frac{\sigma_e B_0^2}{k} u + \frac{\varrho g}{k} = 0. \tag{2.9}$$

As in most problems involving thin films of conducting or non-conducting fluids, we have, in deriving equation (2.9), neglected the effects of surface tension and induced magnetic field, and also assumed that there is no external electric field.

In all that follows, we shall investigate the flow characteristics of the liquid film, by solving equation (2.9) for the two cases: one, when the plate is stationary, and two, when the plate is withdrawn with a constant velocity from a liquid bath.

3. Solution

Considering the case of flow in a thin film of liquid adhering to a stationary plate, there is involved the problem of solving equation (2.9) subject to the conditions

$$u = 0 \quad \text{at} \quad y = 0, \tag{3.1a}$$

and

$$\frac{\mathrm{d}u}{\mathrm{d}y} = 0 \quad \text{at} \quad y = h, \tag{3.1b}$$

where $y = h$ is the equation of free surface. Therefore, we attempt the solution by substituting w for $(\varrho g/k) - m^2 u$, where $m^2 = \sigma_e B_0^2/k$, and the problem reduces to solving the equation

$$n' \left|\frac{\mathrm{d}w}{\mathrm{d}y}\right|^{n'-1} \frac{\mathrm{d}^2w}{\mathrm{d}y^2} + w = 0, \tag{3.2}$$

subject to the boundary conditions

$$w = \frac{\varrho g}{k} \quad \text{at} \quad y = 0, \tag{3.3a}$$

and

$$\frac{\mathrm{d}w}{\mathrm{d}y} = 0 \quad \text{at} \quad y = h. \tag{3.3b}$$

Integrating equation (3.2) once and employing (3.3b) yields

$$\frac{dw}{dy} = -A(w^2 - \alpha^2)^{\frac{1}{n'+1}}, \tag{3.4}$$

where $A = \left\{\frac{m^{2n'}(n'+1)}{2n'}\right\}^{\frac{1}{n'+1}}$ and $\alpha = \frac{\varrho g}{k} - m^2 u_h$, u_h being the velocity at the free surface.

Finally, integrating equation (3.4) and using (3.3a) gives the solution in the form

$$A(h-y) = \frac{n'+1}{2n'\alpha}(w^2-\alpha^2)^{\frac{n'}{n'+1}} F\left[\frac{1}{2}, \frac{n'}{n'+1}; \frac{n'}{n'+1}+1; 1-\frac{w^2}{\alpha^2}\right] \tag{3.5}$$

with

$$Ah = \frac{n'+1}{2n'\alpha}\left(\frac{\varrho^2 g^2}{k^2}-\alpha^2\right)^{\frac{n'}{n'+1}} F\left[\frac{1}{2}, \frac{n'}{n'+1}; \frac{n'}{n'+1}+1; 1-\frac{\varrho^2 g^2}{k^2\alpha^2}\right], \tag{3.6}$$

where the function F is known as Gauss' hypergeometric function, convergent for all n' with $\frac{k^2\alpha^2}{\varrho^2 g^2} > \frac{1}{2}$.

These solutions, furthermore, can be seen to yield the solutions

$$u = \frac{\varrho g}{km^2}\left[1 - \frac{\cosh m(h-y)}{\cosh my}\right], \tag{3.7}$$

and

$$u = \frac{n'}{n'+1}\left(\frac{\varrho g}{k}\right)^{\frac{1}{n'}}\left[h^{\frac{n'+1}{n'}} - (h-y)^{\frac{n'+1}{n'}}\right], \tag{3.8}$$

for the particular cases of conducting Newtonian fluid and non-conducting power-law fluid, respectively.

Non-dimensionalizing equations (3.5) and (3.6) by using the dimensionless parameters

$$\overline{U} = u\left(\frac{\varrho}{g^{n'}k}\right)^{\frac{1}{n'+2}}, \quad Y = y\left(\frac{\varrho^2}{g^{n'-2}k^2}\right)^{\frac{1}{n'+2}}, \quad H = h\left(\frac{\varrho^2}{g^{n'-2}k^2}\right)^{\frac{1}{n'+2}}, \quad \text{and} \quad M = m\left(\frac{k^{n'+3}}{g^2\varrho^{n'+3}}\right)^{\frac{1}{2(n'+2)}}, \tag{3.9}$$

there results

$$\bar{A}(H-Y) = \frac{n'+1}{2n'\bar{\alpha}}(W^2-\bar{\alpha}^2)^{\frac{n'}{n'+1}} F\left[\frac{1}{2}, \frac{n'}{n'+1}; \frac{n'}{n'+1}+1; 1-\frac{W^2}{\bar{\alpha}^2}\right] \tag{3.10}$$

with

$$\bar{A}H = \frac{n'+1}{2n'\bar{\alpha}}(1-\bar{\alpha}^2)^{\frac{n'}{n'+1}} F\left[\frac{1}{2}, \frac{n'}{n'+1}; \frac{n'}{n'+1}+1; 1-\frac{1}{\bar{\alpha}^2}\right], \tag{3.11}$$

where

$$W = \frac{wk}{g}, \quad \bar{\alpha} = \frac{\alpha k}{\varrho g}, \quad \text{and} \quad \bar{A} = \left\{\frac{M^{2n'}(n'+1)}{2n'}\right\}^{1/(n'+1)}. \tag{3.12a, b, c}$$

Eliminating $\bar{\alpha}$ from equations (3.10) and (3.11), W and hence $\overline{U}$ can be obtained as a function of Y, H, $\bar{A}$ and n'.

Next, we discuss the effect of magnetic field on the velocity profile according as the liquid is pseudoplastic ($n' < 1$) or dilatant ($n' > 1$). Therefore, considering the case $n' < 1$, we find that the right-hand -side of equation (3.11) diverges for $\bar{\alpha} = 0$, and this case corresponds to large values of $\bar{A}$ and hence the generalized Hartmann number M. The velocity profile at the free surface, for $\bar{\alpha} = 0$, is then given by

$$\overline{U}_h = \frac{1}{M^2}. \tag{3.13}$$

Consideration of the case $n' > 1$, when $\bar{\alpha} \to 0$, on using equation (3.11), yields

$$\begin{aligned}\bar{A}_{\text{crit.}}H &= \operatorname*{Lt}_{\bar{\alpha}\to 0} \frac{n'+1}{2n'\bar{\alpha}}(1-\bar{\alpha}^2)^{\frac{n'}{n'+1}} F\left[\frac{1}{2}, \frac{n'}{n'+1}; \frac{n'}{n'+1}+1; 1-\frac{1}{\bar{\alpha}^2}\right] \\ &= \frac{n'+1}{2n'} \operatorname*{Lt}_{\bar{\alpha}\to 0} F\left[\frac{1}{2}, 1; \frac{n'}{n'+1}+1; \; 1-\bar{\alpha}^2\right] = \frac{n'+1}{n'-1},\end{aligned} \tag{3.14}$$

leading to

$$\bar{A}_{\text{crit.}} = \frac{n'+1}{(n'-1)H}, \tag{3.15}$$

and equation (3.15) combined with (3.12c) gives

$$M_{\text{crit.}} = \left[\frac{2n'}{n'+1}\left\{\frac{n'+1}{(n'-1)\,H}\right\}^{n'+1}\right]^{\frac{1}{2n'}}. \tag{3.16}$$

Here, it may be mentioned that the following properties of hypergeometric functions, namely,

$$F(a, b; c; z) = (1-z)^{-a}\, F\left(a, c-b; c; \frac{z}{z-1}\right),$$

and

$$F(a, b; c; 1) = \frac{\Gamma(c)\,\Gamma(c-a-b)}{\Gamma(c-a)\,\Gamma(c-b)}, \qquad c-a-b > 0$$

have been utilized to derive equation (3.14).

Further, from equation (3.10), we get

$$\bar{A}(H-Y) \to \bar{A}_{\text{crit.}} H W^{\frac{n'-1}{n'+1}}, \tag{3.17}$$

leading to

$$W \to \left[\frac{\bar{A}_{\text{crit.}} - \bar{A} + \bar{A}(H-Y)}{\bar{A}_{\text{crit.}} H}\right]^{\frac{n'+1}{n'-1}}, \tag{3.18}$$

which, in turn, yields

$$\bar{U} \to \frac{1}{M^2}\left[1 - \left\{\frac{1}{H}\left(1 - \frac{\bar{A}}{\bar{A}_{\text{crit.}}}(1-H+Y)\right)\right\}^{\frac{n'+1}{n'-1}}\right]. \tag{3.19}$$

A simple analysis of (3.19) shows that

$$U = \begin{cases} \dfrac{1}{M^2}\left[1 - \left\{\dfrac{1}{H}\left(1 - \dfrac{\bar{A}}{\bar{A}_{\text{crit.}}}(1-H+Y)\right)\right\}^{\frac{n'+1}{n'-1}}\right], & 0 \leqq Y \leqq H - 1 + \dfrac{\bar{A}_{\text{crit.}}}{\bar{A}}, \\ \dfrac{1}{M^2}, & H - 1 + \dfrac{\bar{A}_{\text{crit.}}}{\bar{A}} \leqq Y \leqq H \end{cases} \tag{3.20}$$

provided that $\bar{A} \geqq \bar{A}_{\text{crit.}}$. This then leads to the formation of a zone

$$H - 1 + \frac{\bar{A}_{\text{crit.}}}{\bar{A}} \leqq Y \leqq H, \tag{3.21}$$

which widens when $\bar{A}(\bar{A} > \bar{A}_{\text{crit.}})$ increases, and in this zone, the dilatant liquid flows like a quasi-solid body with a velocity $\bar{U} = 1/M^2$. Outside this zone, that is, in the zone

$$0 \leqq Y \leqq H - 1 + \frac{\bar{A}_{\text{crit.}}}{\bar{A}}, \tag{3.22}$$

the velocity profile will be given as in (3.20).

Additionally, the problem of investigating the effect of magnetic field for any value of generalized Hartmann number M on velocity profiles, though can be tackled by using equations (3.10) and (3.11), is solved conveniently by using the method of shooting involving predictor — corrector method. Thus, confining ourselves to the case of pseudoplastic liquids ($n' < 1$), we solve equation (2.9), rewritten in the form

$$n' \left|\frac{d\bar{U}}{dY}\right|^{n'-1} \frac{d^2\bar{U}}{dY^2} - M^2\,\bar{U} + 1 = 0, \tag{3.23}$$

with the boundary conditions

$$\bar{U} = 0 \quad \text{at} \quad Y = 0 \quad \text{and} \quad \frac{d\bar{U}}{dY} = 0 \quad \text{at} \quad Y = H, \tag{3.24}$$

by using shooting method with a step-length equal to 0.02. In this method, the condition $d\bar{U}/dY$ at $Y = 0$ is found so as to satisfy the boundary condition at $Y = H$ within an error of the order of 0.5×10^{-12}. $\bar{U}$ is then calculated for different values of Y for a film thickness $H = 1$. The computations were performed on EC-1030 computer, and the velocity profiles are shown in Figs. 1 (a) and (b). It is seen that both the velocity and flow rate decrease with decrease in the power-law index n' and increase in the generalized Hartmann number M.

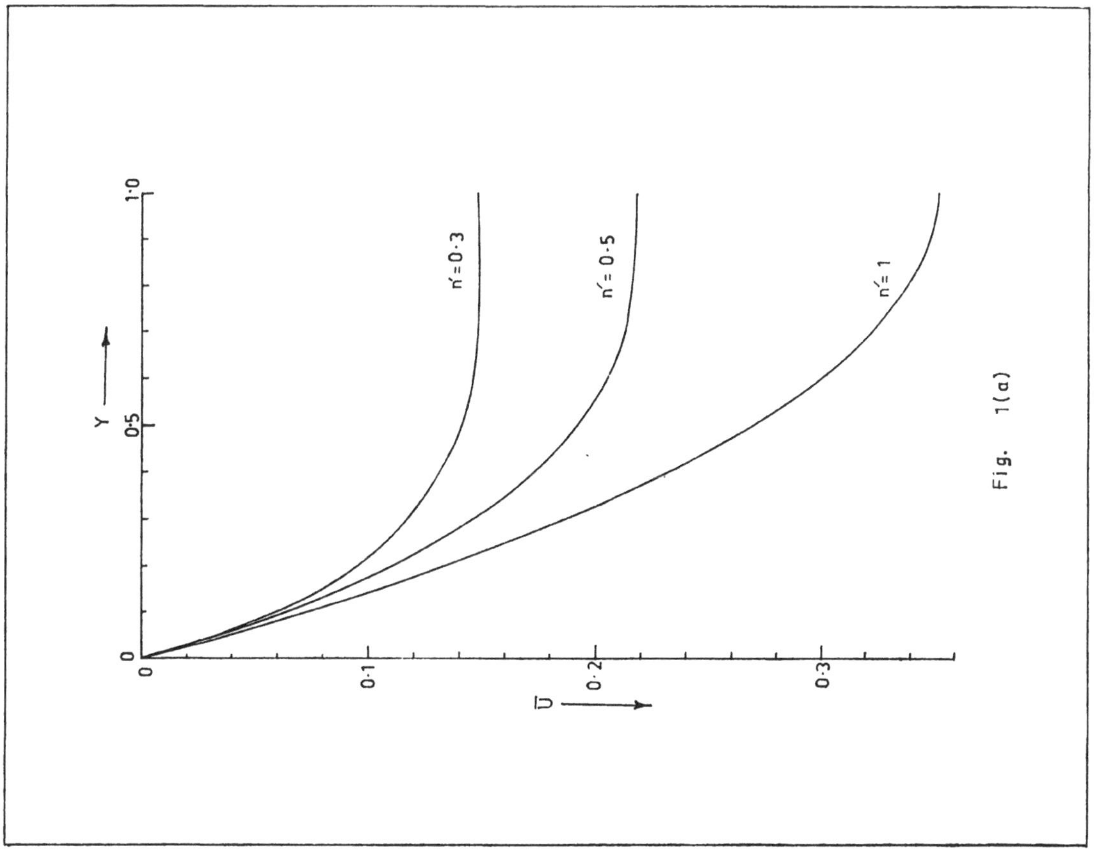

Fig. 1(a). Velocity profiles for a stationary plate: $M = 1$

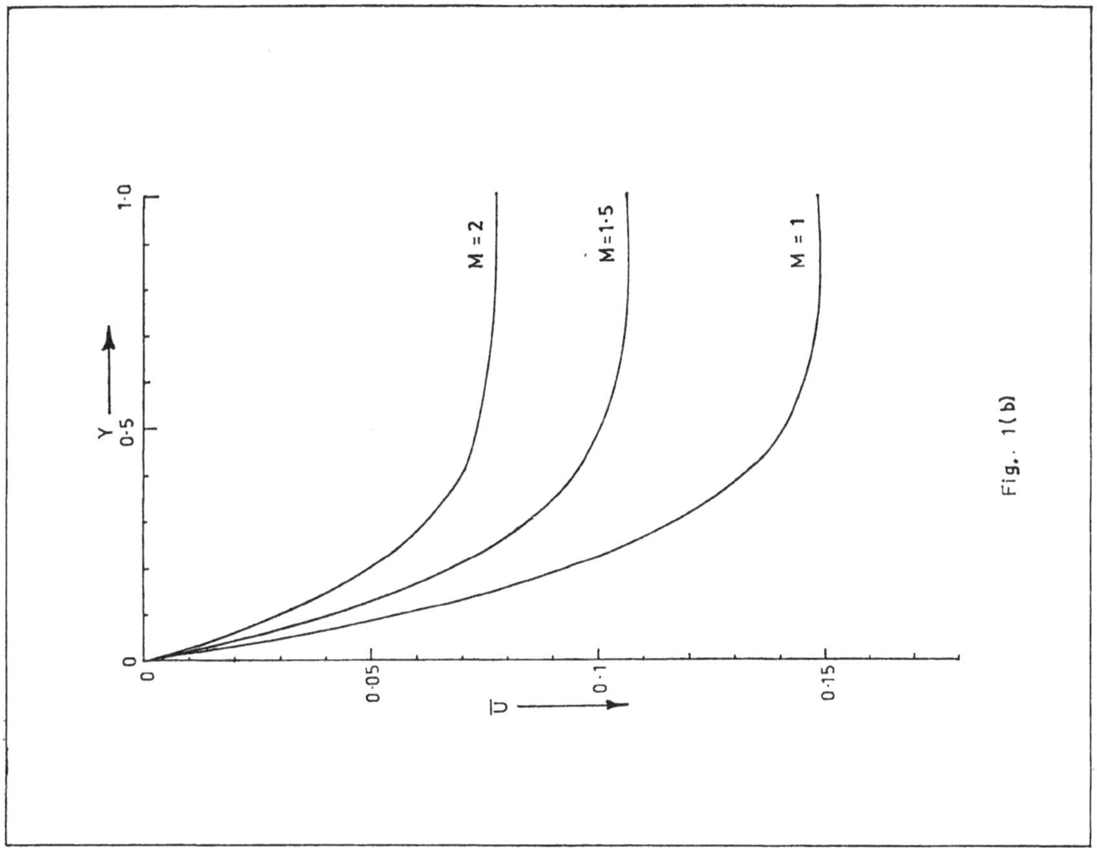

Fig. 1(b). Velocity profiles for a stationary plate: $n' = 0.3$

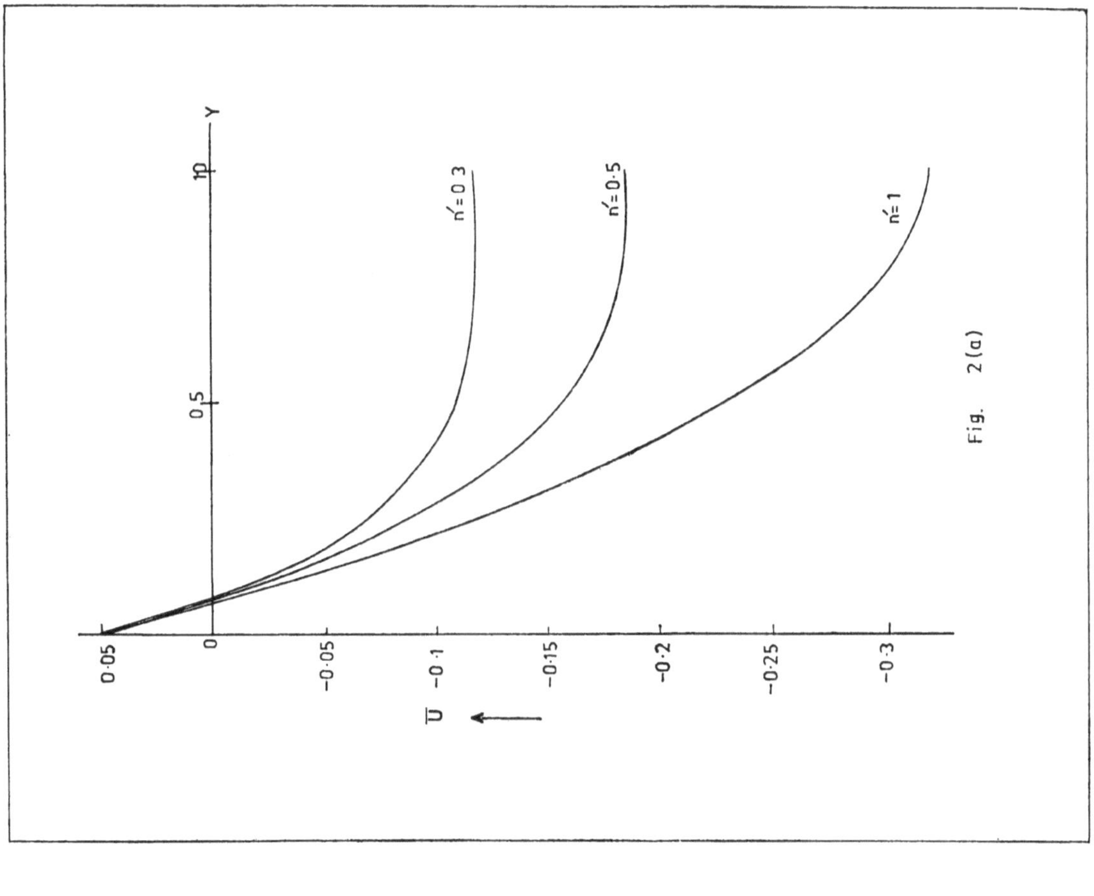

Fig. 2(a). Velocity profiles for a moving plate: $M = 1$

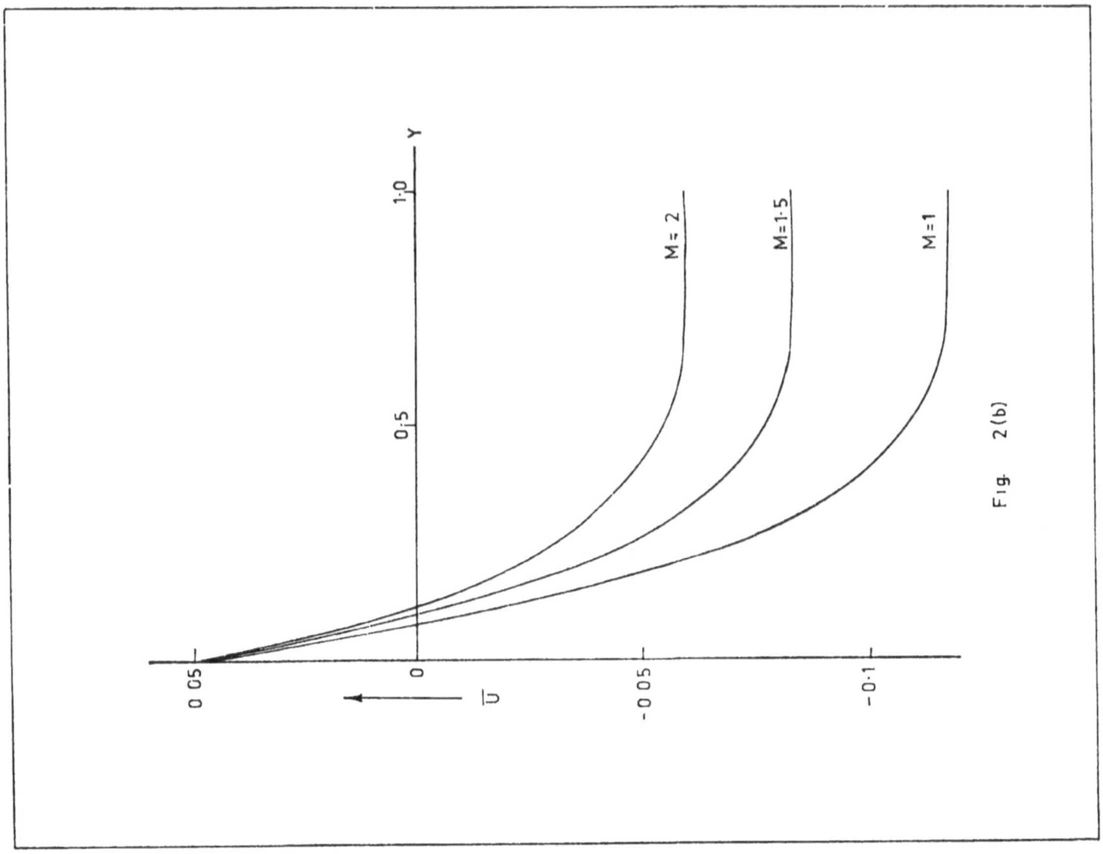

Fig. 2(b). Velocity profiles for a moving plate: $n' = 0.3$

As far as the problem of determining the effect of magnetic field on velocity profiles for the liquid within the thin film adhering to a vertical plane surface withdrawn with a constant velocity from a liquid bath is concerned, we need to solve eq. (2.23), when $\bar{U}$ is changed to $-\bar{U}$ with boundary conditions

$$\bar{U} = -U_W \quad \text{at} \quad Y = 0 \quad \text{and} \quad \frac{d\bar{U}}{dY} = 0 \quad \text{at} \quad Y = H . \tag{3.25}$$

Therefore, once again for the sake of convenience, we use the method of shooting, and determine the velocity profiles exhibited in Figs. (2 a) and (b) for pseudoplastic liquids, when $U_W = 0.05$ and $H = 1$. It is seen from the figures that the phenomenon: the velocity of the liquid first increases and then decreases after the drainage starts, occurs with increase in the generalized HARTMANN number M or with decrease in the power-law index n'. The stagnation point where the fluid velocity vanishes, occurring within the film, is seen to shift away from the plate, as the power-law index n' decreases, or the generalized HARTMANN number M increases. The flow rate is found to decrease with decrease in n' or increase in M.

4. Conclusions

The investigations of the effect of magnetic field on the flow characteristics of a thin film of a power-law conducting liquid adhering to a vertical plane surface carried out in the preceding paragraphs enable us to conclude the following points:

(i) The magnetic field decreases the draining velocity of the liquid, whereas it increases the lifting velocity of the liquid. The same effect is observed when the power-law index decreases.

(ii) The flow rate is decreased with increase in the magnetic field or decrease in the power-law index, and the fall of the film is thereby restricted.

(iii) When a dilatant liquid film flows along a stationary plate, there exists a critical value of the generalized Hartmann number, and as the generalized HARTMANN number exceeds its critical value, there appears a zone near the free surface in which the liquid flows like a quasi-solid body. This zone widens with the increase in the generalized HARTMANN number.

(iv) While withdrawing a vertical plate with a constant velocity from a bath containing a power-law liquid, the stagnation point across a film thickness goes on shifting away from the plate, as the magnetic field increases, or as the power-law index of the liquid decreases.

References

1 MARTINSON, L. K.; PAVLOV, K. B., Unsteady Shear Flows of a Conducting Fluid with a Rheological Power Law, MHD, 7 (1971), 182–189.
2 SARPKAYA, T., Flow of Non-NEWTONian Fluids in a Magnetic Field, A.I.Ch.E.J., 7 (1961), 324–328.
3 MARTINSON, L. K.; PAVLOV, K. B., The Effect of Magnetic Plasticity in Non-NEWTONian Fluids, MHD, 2(3) (1966), 41–44.
4 PAVLOV, K. B., Hydrodynamic Stability of Plane Flows of a Conducting Dilatant Fluid in a Transverse Magnetic Field, MHD, 7 (1971), 454–458.

Eingereicht am 14. 11. 1980

Anschrift: Dr. J. P. AGARWAL, Department of Mathematics, Indian Institute of Technology, Kharagpur — 721302, India

ZAMM 61, 443 – 448 (1981)

V. D. SHARMA / RADHE SHYAM / V. V. MENON

Behaviour of Finite Amplitude Waves in a Radiating Gas

Eine Differentialapproximation für die Gleichungen des Strahlungswärmetransportes in einem grauen Gas von beliebiger Opazität unter Berücksichtigung des Strahlungsflusses, Druckes und der Energiedichte wird angewandt auf die Untersuchung des Einflusses der Wärmestrahlung auf Wachstums- und Zerfallseigenschaften von Wellen endlicher Amplitude und beliebiger Form, die sich in einem thermisch idealen Gas ausbreiten.

A differential approximation for the equations of radiative transfer in a grey gas of arbitrary opacity including the effects of radiative flux, pressure and energy density is applied in a study of the effects of thermal radiation on the growth and decay properties of finite amplitude waves of arbitrary shape propagating through a thermally inviscid gas.

Рассматривается дифференциальная аппроксимация для уравнений транспорта теплоты и излучения в сером газе любой непроницаемости с учетом течения излучения, давления и плотности энергий. Эта аппроксимация применяется для изучения влияния излучения теплоты на свойства роста и распада волн конечной амплитуды и любой формы, которые распространяются в идеальном газе.

1. Introduction

SRINIVASAN and RAM [1] have derived growth equations for the discontinuities in the normal derivatives of flow variables at the wave head propagating through a radiating gas under the optically thin limit; a complete analysis of the solution of their growth equation is performed by SHARMA [2] for plane; cylindrically and spherically expanding waves. The transparent approximation for the radiative transfer equations, used in [1] assumes a grey gas having an absorption coefficient independent of temperature and density, as well as independent of wave length. It may be noted that the constant absorption coefficient assumption is rarely satisfied for high temperature gases; however, the approximation is frequently made in order to obtain the academic solutions. Further, it may be remarked that the neglect of the time derivative in radiative transfer equations suppresses one mode of wave propagation excited by radiation and, therefore, the exact behaviour of waves in radiative gas dynamics is not fully understood as it is evident from their work [1].

The purpose of this paper is to discuss the effects of thermal radiation and the wave front curvature on the growth and decay properties of finite amplitude waves propagating through a thermally radiating inviscid grey gas of arbitrary opacity in three dimensions. Here, we use the differential approximation for radiative transfer, which is valid over the entire optical depth range from the transparent limit to the optically thick limit and the absorption coefficient is taken to be a function of temperature and density; these equations clearly show the existence of radiation induced waves, which have been called precursor radiation by LICK [3] and MOORE [4], and which are followed by modified gas dynamic waves. In the present paper we examine the growth and decay properties of general wave fronts; this situation has not been studied by the workers in radiation gasdynamics till now. Exact predictions of true non-linear progress of the flow variable gradients at the wave head are made for both converging and diverging waves. It is found that the non-linearity in the governing differential equations does not contribute anything to the radiation induced waves; these waves are ultimately damped and the formation of a discontinuous front is not possible from a continuous flow. In the case of modified gas dynamic waves, it is found that all converging compression waves, except in one and only one case in which both the initial principal curvatures are positive and equal, grow without bound only if the initial amplitude of the wave exceeds a critical amplitude. In this special case, it is shown that a compression wave, no matter how weak initially, always grows into a shock before the formation of the focus; we thus characterize converging waves with non-zero initial critical amplitude. In this context, the work on acceleration waves carried out by BOWEN and CHEN [5] and CHEN [6], who characterize diverging waves with non-zero initial critical amplitude, is worth mentioning. However, they also do not refer to the corresponding characterization for converging waves.

2. The Governing Equations

The equations governing the three-dimensional motion of an inviscid gas in which the effects of radiative flux, pressure and energy density may be significant can be written in the form

$$\frac{\partial \varrho}{\partial t} + u_i \varrho_{,i} + \varrho u_{i,i} = 0\,, \tag{1}$$

$$\varrho \frac{\partial u_i}{\partial t} + \varrho u_j u_{i,j} + (p + p_R)_{,i} = 0\,, \tag{2}$$

$$\varrho \frac{\partial}{\partial t} (e_g + (E_R/\varrho)) + \varrho u_i (e_g + (E_R/\varrho))_{,i} + (p + p_R)\, u_{i,i} + q_{i,i} = 0 \tag{3}$$

where u_i are the gas velocity components, p is the gas pressure, ϱ is the gas density, e_g is the internal energy of the gas, E_R is the radiative energy density, p_R is the radiative pressure, q_i are the components of the radiative flux vector, and t is the time. The range of Latin indices is taken to be 1, 2, 3. Here the summation convention on repeated indices is employed, and a comma followed by an index (say i) denotes a partial derivative with respect to a space variable x_i.

Within the differential approximation the equations of radiative transfer may be written as the pair [7]

$$\frac{\partial E_R}{\partial t} + q_{i,i} = -\alpha(cE_R - 4\sigma T^4) , \tag{4}$$

$$\frac{l}{c}\frac{\partial q_i}{\partial t} + \frac{c}{3} E_{R,i} = -\alpha q_i , \tag{5}$$

where α is the absorption coefficient depending on the temperature and density, σ is STEFAN's constant, T is the temperature and c is the velocity of light. Further, within the order of this approximation, the radiative pressure and radiative energy are related according to

$$E_R = 3p_R . \tag{6}$$

Furthermore, the equation of state of the gas is taken to be of the form

$$p = \varrho \mathcal{R} T , \tag{7}$$

where $\mathcal{R}$ is the gas constant together with

$$e_g = p/(\nu - 1)\,\varrho , \tag{8}$$

where ν is the ratio of the specific heats.

Equation (3) with the help of (1), (6) and (8) is conveniently transformed into

$$\frac{\partial p}{\partial t} + u_i p_{,i} + \varrho a_f^2 u_{i,i} + 3(\nu - 1)\left\{\frac{\partial p_R}{\partial t} + u_i p_{R,i}\right\} + (\nu - 1)\, q_{i,i} = 0 , \tag{9}$$

where $a_f = \{(\nu p + 4(\nu - 1)\, p_R)/\varrho\}^{1/2}$ is the effective acoustic speed.

3. Modes of Wave Propagation

We assume that the reader is familiar with the kinematics of moving singular surfaces given, for instance, by TRUESDELL and TOUPIN [8]. Let us consider a moving singular surface Σ, $f(x_i, t) = 0$, across which the flow variables are essentially continuous but discontinuities in their derivatives are permitted, such a singular surface is called a weak wave. The unit normal to the surface is defined by $n_i = f_{,i}/|\mathrm{grad}\, f|$, and the speed of displacement is $G = -(\partial f/\partial t)/|\mathrm{grad}\, f|$. Our convention is that $G > 0$; this corresponds to a singular surface propagating in the direction of n_i. If Z is a function of position and time which is continuous except for a discontinuity at Σ, then the jump of Z is defined by $[Z] = Z_l - Z_0$, where Z_0 is the value of Z immediately ahead of the wave front, and Z_l is the value of Z immediately behind it; it follows that the quantities p, ϱ, q_i, u_i, T, p_R etc. are continuous across Σ, and they will have their subscript '0' values at the wave front.

The normal jump in the velocity gradient $u_{i,j}$ is taken to be the amplitude vector λ_i of the wave, and we set

$$[u_{i,j}]\, n_j = \lambda_i . \tag{10}$$

Forming jumps, across Σ, in equations (1), (2), (4), (5) and (9), and using the first order geometric and kinematic compatibility conditions derived in [8], the definition (10), and the condition that the state ahead of Σ is at rest, we get

$$G\zeta = \varrho_0 \lambda_i n_i , \qquad \varrho_0 G \lambda_i = \xi n_i + \theta n_i , \tag{11, 12}$$

$$3G\theta = \varepsilon_i n_i , \qquad G\varepsilon_i = c^2 \theta n_i , \tag{13, 14}$$

$$G\xi - \varrho_0 a_{f_0}^2 \lambda_i n_i + 3(\nu - 1)\, G\theta - (\nu - 1)\, \varepsilon_i n_i = 0 , \tag{15}$$

where

$$\xi = [p_{,i}]\, n_i , \qquad \zeta = [\varrho_{,i}]\, n_i , \qquad \theta = [p_{R,i}]\, n_i , \qquad \varepsilon_i = [q_{i,j}]\, n_j$$

are the quantities defined on Σ.

It is interesting to note, from (12), that for the type of material we have considered here, only longitudinal waves are possible. Equations (11) to (15) constitute a set of nine homogeneous equations in nine unknowns λ_i, ε_i, ξ, ζ and θ. Hence the necessary condition for the existence of non-trivial solutions of (11) to (15) is that the determinant of the coefficients matrix must vanish. This yields

$$G = 0 , \qquad G = \pm\, c/3^{1/2} , \qquad G = \pm\, a_{f_0} .$$

Here we are interested in situations when $G \neq 0$, which means that the wave surface is not a material surface. Further, for an advancing wave surface, we shall take G to be positive. Thus, we find that there are two waves present in the gas, one which propagates with the speed

$$G = c/3^{1/2} \tag{16}$$

has attributes which are basically due to radiation and it may be identified as radiation induced wave. The other, which propagates with the speed

$$G = a_{f_0} \tag{17}$$

is essentially a modified gas dynamic wave.

4. Behaviour at the Wave Head

(i) Radiation Induced Waves

Let us consider discontinuities in the first derivatives of flow quantities propagating along a radiation induced wave Σ. The flow ahead of this wave remains undisturbed and it is convenient to take it to be of uniform state; the equations (11) to (15), using (16), yield

$$\theta = \varepsilon/3^{1/2}c\,, \qquad \zeta = 3^{1/2}\varepsilon/c^3\,\{1 - (3a_{f_0}^2/c^2)\}\,, \tag{18), (19}$$

$$\xi = 3^{1/2}a_{f_0}^2\varepsilon/c^3\{1 - (3a_{f_0}^2/c^2)\}\,, \qquad \lambda = \varepsilon/\varrho_0 c^2\{1 - (3a_{f_0}^2/c^2)\}\,, \tag{20), (21}$$

where $\varepsilon = \varepsilon_i n_i$ and $\lambda = \lambda_i n_i$.

If we differentiate (7) with respect to x_i, form jumps across Σ, multiply the resulting equation by n_i, and make use of the relations (19) and (20), we obtain

$$[T_{,i}]\,n_i = \frac{3^{1/2}(a_{f_0}^2 - a_{T_0}^2)\,\varepsilon}{\mathcal{R}\varrho_0 c^3\{1 - (3a_{f_0}^2/c^2)\}} = \chi \tag{22}$$

(say), where a_{T_0} is the isothermal speed of sound given by

$$a_{T_0}^2 = p_0/\varrho_0\,.$$

When (5) is differentiated partially with respect to t and use is made of (4), we get

$$q_{i,ij} - \frac{3}{c^2}\frac{\partial^2 q_j}{\partial t^2} - \frac{6\alpha}{c}\frac{\partial q_j}{\partial t} - 3\alpha^2 q_j - 16\alpha\sigma T^3 T_{,j} - \frac{3q_j}{c}\left\{\frac{\partial\alpha}{\partial\varrho}\frac{\partial\varrho}{\partial t} + \frac{\partial\alpha}{\partial T}\frac{\partial T}{\partial t}\right\} +$$

$$+ (cE_R - 4\sigma T^4)\left(\frac{\partial\alpha}{\partial\varrho}\varrho_{,j} + \frac{\partial\alpha}{\partial T}T_{,j}\right) = 0\,.$$

If we multiply this equation by n_j and form jumps across Σ, we find, on using the second order geometrical and kinematical conditions derived in [8] and the relations (19), (21) and (22) that

$$\frac{\delta\lambda}{\delta t} = -(\Lambda_1 - \Lambda_2)\,\lambda\,, \tag{23}$$

where $\delta/\delta t$ is the THOMAS [9] time derivative along an orthogonal trajectory of Σ, $\Lambda_1 = \{\alpha - (\Omega/3^{1/2})\}\,c$ and

$$\Lambda_2 = \frac{\{1 - (3a_{f_0}^2/c^2)\}^{-1}}{2\mathcal{R}\varrho_0 c^2}\left[16\alpha\sigma T_0^3(a_{f_0}^2 - a_{T_0}^2) - (\sqrt{3}\,q_n + cE_R - 4\sigma T^4)\left\{\left(\frac{\partial\alpha}{\partial\varrho}\right)_0\varrho_0\mathcal{R} + \left(\frac{\partial\alpha}{\partial T}\right)_0(a_{f_0}^2 - a_{T_0}^2)\right\}\right].$$

Ω and q_n are, respectively, the mean curvature of Σ and the normal component of the flux vector defined as $\Omega = \frac{1}{2}\,g^{\alpha\beta}b_{\alpha\beta}$ and $q_n = q_i n_i$ with $g^{\alpha\beta}$ and $b_{\alpha\beta}$ being the first and second fundamental forms of Σ; the range of Greek letter indices, which refer to the surface coordinates y^α, is taken to be 1, 2.

Equation (23) is the governing differential equation of the amplitude of a radiation induced wave. The behaviour of the solution of (23) can, of course, be readily established. The mean curvature Ω at any point of the wave surface Σ has the representation [10]

$$\Omega = (\Omega_0 - K_0 G t)/(1 - 2\Omega_0 G t + K_0 G^2 t^2)\,, \tag{24}$$

where $\Omega_0 = (k_1 + k_2)/2$ and $K_0 = k_1 k_2$ are, respectively, the mean and Gaussian curvatures of Σ at $t = 0$ with k_1 and k_2 being the principal curvatures, and G is the speed of propagation of the wave surface which for the present case is equal to $c/3^{1/2}$. When k_1 and k_2 are both non-positive, the wave is divergent. On the other hand if one or both the principal curvatures are positive then it corresponds to the case of a convergent wave. Here, we shall consider these two cases.

Equation (23), after substitution from (24), can be integrated to yield

$$\lambda = \lambda_0 I \exp(-\alpha c t)\,, \tag{25}$$

where $I = \{(1 - (k_1 c t/3^{1/2}))\,(1 - (k_2 c t/3^{1/2}))\}^{-1/2}$, and λ_0 is the value of λ at $t = 0$. In (23) the term Λ_2 which contains c^2 in the denominator, has been neglected in comparison with Λ_1 in presenting the result (25).

For diverging waves, it is apparent from (25) that $\lambda \to 0$ as $t \to \infty$. This means that the radiation induced waves are damped and formation of a front, carrying discontinuities in the flow quantities, is not possible from a continuous flow. From the expressions (18) to (22), it follows that the quantities λ, ξ, ζ and χ are small compared with ε and θ as it should be in radiation induced waves.

For converging waves, there exists a finite time t^* given by the smallest positive root of

$$(3^{1/2} - k_1 c t)\,(3^{1/2} - k_2 c t) = 0\,,$$

such that $\lambda \to \infty$ as $t \to t^*$; this corresponds to the formation of a focus.

(ii) Modified Gas Dynamic Waves

We shall now consider discontinuities in the first derivatives of flow quantities propagating along a modified gas dynamic wave Σ. In general, the flow ahead of this wave will be disturbed by radiation induced waves. But we have just seen that for radiation induced waves λ, ξ, ζ and χ are small compared to ε and θ and hence the main effects can be seen for the special case in which $u_i = 0$ and p, ϱ and T are constant ahead of the wave Σ. For a modified gas dynamic wave, equations (11) to (15), using (17), yield

$$a_{f_0}\zeta = \varrho_0\lambda\,, \qquad \varrho_0 a_{f_0}\lambda = \xi\,, \qquad \varepsilon = \theta = 0\,. \tag{26, 27, 28}$$

If we differentiate (2), (4), (5) and (9) partially with respect to x_k, take jumps across Σ, and multiply the resulting equations by n_k, then, using the compatibility conditions derived in [8], we obtain

$$\varrho_0 \frac{\delta\lambda}{\delta t} - \varrho_0 a_{f_0}\bar{\lambda} + \bar{\xi} + \bar{\theta} = 0\,, \tag{29}$$

$$\bar{\varepsilon} - 3a_{f_0}\bar{\theta} - \frac{16\alpha\sigma T_0^3}{\mathcal{R} a_{f_0}}(a_{f_0}^2 - a_{T_0}^2)\,\lambda + (cE_{R_0} - 4\sigma T_0^4)\left\{\left(\frac{\partial\alpha}{\partial\varrho}\right)_0 \zeta + \left(\frac{\partial\alpha}{\partial T}\right)_0 \chi\right\} = 0\,, \tag{30}$$

$$c^2\bar{\theta} - a_{f_0}\bar{\varepsilon} - q_n\left\{\left(\frac{\partial\alpha}{\partial\varrho}\right)_0 \zeta + \left(\frac{\partial\alpha}{\partial T}\right)_0 \chi\right\} = 0\,, \tag{31}$$

$$\frac{\partial\xi}{\partial t} - a_{f_0}(\bar{\xi} - \varrho_0 a_{f_0}\bar{\lambda}) + (\nu + 1)\,\xi\lambda - 2\varrho_0 a_{f_0}^2 \Omega\lambda + (\nu - 1)\,(\bar{\varepsilon} - 3a_{f_0}\bar{\theta}) = 0\,, \tag{32}$$

where

$$\bar{\lambda} = [u_{i,jk}]\, n_i n_j n_k\,, \qquad \bar{\xi} = [p_{,ij}]\, n_i n_j\,, \qquad \bar{\theta} = [p_{R,ij}]\, n_i n_j\,, \qquad \bar{\varepsilon} = [q_{i,jk}]\, n_i n_j n_k$$

are the quantities defined on Σ, and Ω is the mean curvature of Σ given by (24).

Using (26) to (28) and (30) to (32) in (29), we obtain

$$\frac{\delta\lambda}{\delta t} + (\mu_0 - a_{f_0}\Omega)\,\lambda + \frac{(\nu + 1)}{2}\lambda^2 = 0\,, \tag{33}$$

where

$$\mu_0 = \frac{(\nu - 1)}{2\mathcal{R}\varrho_0 a_{f_0}^2}\left\{16\alpha\sigma T_0^3(a_{f_0}^2 - a_{T_0}^2) - \left(\mathcal{R}\varrho_0\left(\frac{\partial\alpha}{\partial\varrho}\right)_0 + \left(\frac{\partial\alpha}{\partial T}\right)_0 (a_{f_0}^2 - a_{T_0}^2)\right)(cE_{R_0} - 4\sigma T_0^4)\right\} +$$
$$+ \frac{1}{2\mathcal{R}\varrho_0 a_{f_0}^2 c^2\left(1 - \dfrac{3a_{f_0}^2}{c^2}\right)}\left\{16\alpha\sigma T_0^3 a_{f_0}(a_{f_0}^2 - a_{T_0}^2) - \left(\mathcal{R}\varrho_0\left(\frac{\partial\alpha}{\partial\varrho}\right)_0 + \right.\right.$$
$$\left.\left. + \left(\frac{\partial\alpha}{\partial T}\right)_0 (a_{f_0}^2 - a_{T_0}^2)\right)(q_n + a_{f_0}(cE_{R_0} - 4\sigma T_0^4))\right\}.$$

Equation (33) is the governing differential equation of the amplitude of a modified gas dynamic wave. In the following discussion we shall see how nicely this equation shows the interplay of damping and steepening tendencies of the linear damping term $(\mu_0 - a_{f_0}\Omega)\,\lambda$ and the non-linear steepening term $\left(\frac{\nu + 1}{2}\right)\lambda^2$. The solution of (33) is

$$\lambda = \frac{\lambda_0 \exp(-\mu_0 t)\, I_1(t)}{1 + \lambda_0 \dfrac{(\nu + 1)}{2} I_2(t)}\,, \tag{34}$$

where λ_0 is the value of λ at the wave front at $t = 0$,

$$I_1(t) = \{(1 - k_1 a_{f_0} t)\,(1 - k_2 a_{f_0} t)\}^{-1/2}\,, \tag{35}$$

and

$$I_2(t) = \int_0^t I_1(\hat{t}) \exp(-\mu_0 \hat{t})\, d\hat{t}\,. \tag{36}$$

It is clear from (34) that the behaviour of the wave amplitude will depend critically on the sign of μ_0.

The analysis of the solution of (33) is similar to that given by [5] except for the case of converging waves with $k_1 = k_2 > 0$, which has not been considered till now. Our main purpose is to discuss this situation but for the sake of completeness we prefer here to discuss diverging waves also in brief.

Case (i). Diverging Waves (k_1, $k_2 \leqq 0$): When $\mu_0 > 0$, it is clear from (36) that the integral $I_2(t)$ converges to a finite limit as $t \to \infty$. Hence, it follows from (34) that if $\lambda_0 > 0$ (i.e. an expansion wave front) then $\lambda \to 0$ as $t \to \infty$ and the wave decays and damps out ultimately. But if $\lambda_0 < 0$ (i.e. a compression wave front) then there exists a positive critical value λ_c of the wave amplitude given by

$$\lambda_c = \left\{ \frac{(\nu+1)}{2} \int_0^t I_1(\tau) \exp(-\mu_0 \tau) \, d\tau \right\}^{-1} \tag{37}$$

such that waves with initial amplitude $|\lambda_0| < \lambda_c$ damp to zero ($\lambda \to 0$ as $t \to \infty$) and the waves with initial amplitude $|\lambda_0| > \lambda_c$ grow without bound in a finite time $t_c (|\lambda| \to \infty$ as $t \to t_c)$ given by

$$I_2(t_c) = 2/(\nu + 1) \, |\lambda_0| \, . \tag{38}$$

Further, if $\lambda_0 < 0$ and $|\lambda_0| = \lambda_c$ then, on using L'HOSPITAL and LEIBNIZ rules, it follows from (34) that $|\lambda| \to \to 2\mu_0(\nu + 1)^{-1}$ as $t \to \infty$. This means that a compression wave with $|\lambda_0| = \lambda_c$ can neither terminate into a shock wave nor can it ever completely damp out. From (37) it follows that $(\partial\lambda_c/\partial\mu_0) > 0$ which means that the critical value of the initially discontinuity increases with μ_0. Hence, the effects of thermal radiation with $\mu_0 > 0$ are to delay the onset of a shock wave. Also $(\partial\lambda_c/\partial|k_1|) > 0$ and $(\partial\lambda_c/\partial|k_2|) > 0$ which imply that the initial curvatures have a stabilizing effect on the tendency of the wave surface to grow into a shock in the sense that an increase in the value of initial curvature causes an increase in the critical value λ_c.

When $\mu_0 < 0$, it follows from (36) that $\lim_{t\to\infty} I_2(t) = \infty$. Therefore, using L'HOSPITAL's rule, it follows from (34) that for $\lambda_0 > 0$, $\lambda \to 2|\mu_0| (\nu + 1)^{-1}$ as $t \to \infty$ (i.e. all expansion waves ultimately take a stable wave form). But if $\lambda_0 < 0$ then $|\lambda| \to \infty$ as $t \to \bar{t}_c$ where $\bar{t}_c$ is given by the solution of

$$\int_0^{\bar{t}_c} I_1(\tau) \exp(|\mu_0| \, \tau) \, d\tau = 2/(\nu + 1) \, |\lambda_0| \, .$$

Thus when $\mu_0 < 0$, all compression waves, no matter how small be their initial amplitude, always terminate in a shock wave. Further, in this case, $(\partial\bar{t}_c/\partial|\mu_0|) < 0$ and $(\partial\bar{t}_c/\partial|k_i|) > 0$ $(i = 1, 2)$ i.e., $\bar{t}_c$ is a decreasing function of $|\mu_0|$ but an increasing function of k_i. Thus, we reach to an important conclusion that *the thermal radiation effects with $\mu_0 < 0$ are to swift the onset of a shock wave.*

Case (ii). Converging Waves: In the analysis of converging waves, for which at least one of the initial principal curvatures is positive, the integral

$$I_2(t^*) = \int_0^t [(1 - k_1 a_{f_0} t)(1 - k_2 a_{f_0} t)]^{-1/2} \exp(-\mu_0 t) \, dt \tag{39}$$

plays a crucial role, where t^* is the smallest positive root of the equation

$$(1 - k_1 a_{f_0} t)(1 - k_2 a_{f_0} t) = 0 \, . \tag{40}$$

Surprisingly, CHEN [6] (§ 3.4, p. 44) says "We have no idea as to whether the integral (39) is finite or infinite". BOWEN and CHEN [5], however, have done better by making the statement "Observe, also, from (2.18) that for a converging wave we can have $\mu_0 \leqq 0$ and still have a non-zero critical initial amplitude". The analysis in [5] and [6] would have been more satisfactory if the convergence of the integral $I_2(t^*)$ was also considered at the same place.

One can easily show that the integral $I_2(t^*)$ is infinite if, and only if, both k_1 and k_2 are positive and equal i.e. $k_1 = k_2 > 0$. For if $k_1 = k_2 > 0$, by substituting $z = t^* - t$, we find that the singularity as $t \to t^*$ in the integral of $I_2(t^*)$ is of the form $z^{-1}g(z)$ as $z \to 0$, where $g(z)$ is both bounded and bounded away from zero, and the function z^{-1} is not integrable on any interval $[0, \delta]$, $\delta > 0$. If $k_1 \neq k_2$ and at least one of k_1 and k_2 is positive, then the integral $I_2(t^*)$ is finite; this follows from the argument that the singularity as $t \to t^*$ in the integral of $I_2(t^*)$ is of the type of $z^{-1/2}h(z)$ as $z \to 0$, where the function $h(z)$ is again bounded and bounded away from zero, and the function $z^{-1/2}$ is integrable over every interval $[0, \delta]$, $\delta > 0$. If $k_1 \leqq 0$ and $k_2 \leqq 0$, the integral $I_2(t^*)$ is obviously finite.

Thus, it follows from (34) that when $k_1 \neq k_2$ and at least one of k_1 and k_2 is positive then, irrespective of the sign of μ_0, there exists a finite t^* given by the smallest positive root of (40) such that for $\lambda_0 > 0$, $|\lambda| \to \infty$ as $t \to t^*$, i.e., a focus is formed. But if $\lambda_0 < 0$ then there exists a critical amplitude $\hat{\lambda}_c$ given by

$$\hat{\lambda}_c = 2/(\nu + 1) \, I_2(t^*) \tag{41}$$

such that if $|\lambda_0| < \hat{\lambda}_c$ then $|\lambda| \to \infty$ as $t \to t^*$, which corresponds to the formation of a focus but not the shock. On the other hand if $|\lambda_0| > \hat{\lambda}_c$ then $|\lambda| \to \infty$ as $t \to \hat{t}_c$ where $\hat{t}_c$ is given by

$$I_2(\hat{t}_c) = 2/(\nu + 1) \, |\lambda_0| \, .$$

It is evident that $\hat{t}_c < t^*$. Thus, when $|\lambda_0| > \hat{\lambda}_c$, we find that the shock wave is formed before the focus can. The case $|\lambda_0| = \hat{\lambda}_c$ corresponds to the simultaneous formation of a shock and a focus.

BOWEN and CHEN [5] have shown that not all converging waves will grow into shock waves. However, they have ignored the possibility $k_1 = k_2 > 0$ in their analysis. In this special case, and only in this case, $I_2(t) \to \infty$ as $t \to t^*$, and thus the critical initial amplitude $\hat{\lambda}_c$ given by (41) vanishes. In this case, it follows from (34) that, irrespective of the sign of μ_0, there exists a finite time $\tilde{t} > 0$ such that $\tilde{t} < t^*$, $I_2(\tilde{t}) = 2/(\nu + 1) \, |\lambda_0|$ and $\lim_{t\to\tilde{t}} |\lambda| = \infty$. This means that when and only when $k_1 = k_2 > 0$, all converging compression waves, irrespective of the sign of μ_0, will grow into shock waves before the formation of the focus, no matter how small be their initial amplitude. We thus characterize converging waves with non-zero initial critical amplitude.

References

1 SRINIVASAN, S., and RAM, R., Propagation of sonic waves in radiating gases, ZAMM **57**, 191—193 (1977).
2 SHARMA, V. D., A note on the propagation of sonic waves in radiating gases, ZAMM **60**, 116—117 (1980).
3 LICK, W. J., Propagation of small disturbances in a radiating gas, J. Fluid Mech. **18**, 274—284 (1964).
4 MOORE, F. K., Effect of radiative transfer on a sound wave traveling in a gas having ν near one, Phys. Fluids **9**, 70—80 (1966).
5 BOWEN, R. M., and CHEN, P. J., Some comments on the behaviour of acceleration waves of arbitrary shape, J. Math. Phys. **13**, 948—950 (1972).
6 CHEN, P. J., Selected Topics in Wave Propagation, Noordhoff Int. Publ. p. 44 (1976).
7 VINCENTI, W. G., and KRUGER, C. H., Introduction to Physical Gas Dynamics, Wiley (1965).
8 TRUESDELL, C., and TOUPIN, R. A., The classical Field Theories, Handbuch der Physik, Band III/1 (Ed. S. FLÜGGE), Chap. C. Berlin-Göttingen-Heidelberg Springer (1960).
9 THOMAS, T. Y., Extended Compatibility Conditions for the Study of Surfaces of Discontinuity in Continuous Mechanics, J. Math. Mech. **6**, 311—322 (1957).
10 LANE, E. P., Metric Differential Geometry of Curves and Surfaces, Chicago Univ. Press, p. 209 (1940).

Eingereicht am 22. 4. 1980

Anschrift: Dr. V. D. SHARMA, Applied Mathematics Section, J.T., B.H.U. Varanasi 221005, India

ZAMM 61, 449 – 459 (1981)

B. FUGLEDE

On a Direct Method of Integral Equations for Solving the Biharmonic Dirichlet Problem

Untersucht wird eine Variante der Integralgleichungsmethode zur numerischen Lösung des Dirichlet-Problems der Laplace-Gleichung oder der biharmonischen Gleichung in einem beschränkten ebenen Gebiet. Diese Methode beruht auf der dritten Greenschen Identität; ihr Hauptvorteil liegt darin, daß sie von den vorgegebenen Dirichletschen Randwerten direkt zu den physikalisch interessanten Neumannschen Randwerten führt. Unlängst wurde von S. Christiansen und P. Hougaard vermutet, daß die betrachtete(n) Integralgleichung(en) für gewisse Werte des äußeren Abbildungsradius des Gebietes noch andere Lösungen als die des ursprünglichen Dirichlet-Problems hat (haben). Sie bewiesen das auch für das Kreisgebiet und fanden in anderen Fällen numerische Anzeichen für die Richtigkeit der Vermutung. In diesem Artikel werden wir die genannte und damit verwandte Vermutungen für ein allgemeines Gebiet mit glattem Rand beweisen. Numerische Fragen der betrachteten Methode werden nicht diskutiert.

In recent years an alternative method of integral equations has been used for numerical solution of the Dirichlet problem for the Laplace equation or the biharmonic equation in a bounded plane domain. This method is based simply on Green's third identity, and its main advantage is that is leads directly from the prescribed Dirichlet boundary data to the (physically relevant) Neumann data. It has been conjectured recently by S. Christiansen and P. Hougaard that, for certain values of the outer mapping radius of the domain, the integral equation(s) in question has (have) other solutions than that which stems from the solution to the Dirichlet problem. They proved this for a circular domain and gave numerical indications in another case. In the present paper we prove this and related conjectures in the case of a general domain with smooth boundary. We do not discuss the numerical aspects of the method in question.

Рассмотрим вариант метода интегральных уравнений для численного решения задачи Дирихле для уравнения Лапласа либо бигармонического уравнения в конечной плоской области. Метод основан на использовании третьей формулы Грина и имеет преимущество вести от заданных первых краевых данных прямо к физически интересным вторым краевым данным. Недавно С. Христианзен и П. Хаугард высказали предположение, что интегральное уравнение (либо уравнения) для некоторых значений радиуса внешнего отображения области имеет еще и другие решения, кроме решения первоначальной задачи Дирихле. Они доказали это для круговой области и получили в других случаях численные результаты, указывающие на правильность предположения. В настоящей статье докажем названное и сходные предположения для общей области с гладкой границей. Вопросы численной реализации рассмотренного метода не обсуждаем.

Introduction

The classical method of integral equations for solving the Dirichlet problem — say for the Laplace equation — was designed to lead to an equivalent integral equation of the second kind on the boundary Γ of the bounded plane domain D in question. The unknown function in the integral equation was the moment density generating the double layer potential by which the solution should be represented.

When it comes to numerical solution of the Dirichlet problem this method has the disadvantage that it does not lead directly to the physically relevant unknown function(s) viz. the Neumann data of the solution u (that is, the normal derivative $\partial u/\partial \nu$ in the case of the Laplace equation). An additional integration process is needed to pass from the moment density to the Neumann data of the solution u.

Accordingly, in recent years, several authors have preferred to use another method of integral equations serving to lead directly from the prescribed Dirichlet data (the boundary values of u in the Laplacian case) to the Neumann data. This direct method is conceptually simpler than the classical method (especially for higher order equations). It amounts simply to writing down Green's 3rd identity for the solution u and taking into account the jump in the double layer potential (entering also here) at the boundary. On the other hand, it leads to an integral equation of the *first* kind and hence is not suited for existence proofs for the Dirichlet problem, but such proofs are of course available by other methods.

In the case of the Laplace equation in a plane domain D it is known that the integral equation obtained by this "direct" method has, for a certain value of the outer mapping radius of D (the value 1 if the fundamental solution to the Laplace equation is chosen as $-\log r$), other solutions than the relevant solution (i.e. the solution stemming from the unique solution to the Dirichlet problem). This is easy to prove, see [5], or § 2 below.

Recently the same phenomenon has been observed by S. CHRISTIANSEN and P. HOUGAARD [2] in the case of the biharmonic Dirichlet problem in the plane. In the case of circular domains they proved that the pair of integral equations (obtained by the direct method) has irrelevant solutions when the radius of the circle is 1 or e^{-1}. They conjectured that a similar non-uniqueness occurs for a general plane domain D when, and only when, the outer mapping radius d of D equals 1 or e^{-1}, and they gave numerical indications for this in the case of ellipses for which the ratio of the axes is 2:1.

In the present paper we shall prove this conjecture (and others from [2]) in the case of a general plane domain D (with sufficiently smooth boundary Γ). We also prove that, in dimensions $n > 2$, there is always uniqueness, that is, the pair of integral equations obtained by the direct method has just one solution (the one which stems from the solution of the biharmonic Dirichlet problem).

1. Formulation of the results

Let D denote a bounded connected open set in $\mathbb{R}^n$ $(n \geqq 2)$ with boundary Γ such that Γ (more precisely: D together with Γ) is of class $C^{1+\alpha}$ (also denoted by $C^{1,\alpha}$) for some $\alpha > 0$, cf. [3, § 6.2, § 4.1]. The complement of $\bar{D} = D \cup \Gamma$ consists of an unbounded domain D_e with boundary Γ_e and (possibly) a finite number of bounded components, the union of which is denoted by D_i and its boundary by Γ_i. Writing $E = \complement \bar{D}$, we thus have

$$E = D_e \cup D_i\,, \qquad \Gamma = \Gamma_e \cup \Gamma_i\,.$$

The possible presence of an interior boundary Γ_i turns out to be of no influence for the problem under discussion.

The *biharmonic Dirichlet problem* consists in finding a function $u \in C^1(\bar{D}) \cup C^4(D)$ such that $\Delta u \in C^1(\bar{D})$ and[1])

$$\text{(I)} \qquad \begin{cases} \Delta^2 u = 0 & \text{in } D\,, \\ u = f_0 \ \text{ and } \ \partial u/\partial\nu = f_1 & \text{on } \Gamma\,, \end{cases}$$

where f_0 and f_1 are given functions on Γ of classes C^1 and C^0 (continuous), respectively, and ν denotes the outward unit normal to Γ (pointing from D into E).

The associated pair of integral equations depends on the choice of a particular fundamental solution G for the operator $-\Delta^2$, and likewise a fundamental solution g for the operator $-\Delta$. Following [2], we take (in the plane case)

$$G(x, y) = \frac{1}{8\pi^2}\,|x-y|^2 \log\frac{1}{|x-y|}\,, \qquad g(x, y) = \frac{1}{2\pi}\log\frac{1}{|x-y|}\,.$$

Another natural choice of G is discussed in § 9 below. For the case of dimension $n > 2$, see § 10.

Green's 3rd identity, applied to a function $u \in C^1(\bar{D})$ which is biharmonic in D, and such that $\Delta u \in C^1(\bar{D})$, gives

$$\int_\Gamma \left(g\frac{\partial u}{\partial\nu} - \frac{\partial g}{\partial\nu}u\right) ds = \int_D g\Delta u + \begin{cases} u & \text{in } D \\ \frac{1}{2}u & \text{on } \Gamma\,. \\ 0 & \text{in } E \end{cases}$$

Here $g = g(x, y)$, $u = u(y)$ and $\Delta u = \Delta u(y)$ under the integral signs, while $u = u(x)$ in the second term on the right, and ds refers to the arc length on Γ. Green's 2nd identity, applied to $G(x, .)$ and Δu, gives similarly

$$\int_\Gamma \left(G\frac{\partial\Delta u}{\partial\nu} - \frac{\partial G}{\partial\nu}\Delta u\right) ds = -\int_D \Delta G\,\Delta u\,.$$

Adding these two equations, and inserting $\Delta G = g - (1/2\pi)$, we obtain what may be termed Green's 3rd identity for the equation $\Delta^2 u = 0$ in D:

$$\int_\Gamma \left(G\frac{\partial\Delta u}{\partial\nu} - \frac{\partial G}{\partial\nu}\Delta u + \Delta G\frac{\partial u}{\partial\nu} - \frac{\partial\Delta G}{\partial\nu}u\right) ds = \begin{cases} u & \text{in } D \\ \frac{1}{2}u & \text{on } \Gamma, \\ 0 & \text{in } E \end{cases}$$

because $\int_\Gamma (\partial u/\partial\nu)\,ds = \int_D \Delta u$. Since $\Delta u \in C^1(\bar{D})$, and Δu is harmonic in D, we further have, by Green's 3rd identity for the Laplace equation in D:

$$\int_\Gamma \left(g\frac{\partial\Delta u}{\partial\nu} - \frac{\partial g}{\partial\nu}\Delta u\right) ds = \begin{cases} \Delta u & \text{in } D \\ \frac{1}{2}\Delta u & \text{on } \Gamma. \\ 0 & \text{in } E \end{cases}$$

Altogether we are thus led to the following pair of integral equations on Γ with unknown functions v and w (replacing Δu and $\partial\Delta u/\partial\nu$, respectively):

$$\text{(II)} \qquad \begin{cases} \displaystyle\int_\Gamma \left\{G(x, y)\,w(y) - \frac{\partial G(x, y)}{\partial\nu(y)}\,v(y) + \Delta G(x, y)\,f_1(y) - \frac{\partial\Delta G(x, y)}{\partial\nu(y)}\,f_0(y)\right\} ds(y) = \frac{1}{2}f_0(x)\,, \\ \displaystyle\int_\Gamma \left\{g(x, y)\,w(y) - \frac{\partial g(x, y)}{\partial\nu(y)}\,v(y)\right\} ds(y) = \frac{1}{2}v(x)\,. \end{cases}$$

---

[1]) The condition $\Delta u \in C^1(\bar{D})$ is understood to mean that Δu extends from D to a function of class $C^1(\bar{D})$ (likewise denoted by Δu). By imposing this condition we ensure uniqueness. In fact, if u is a solution (as stated) of the homogeneous biharmonic Dirichlet problem ($f_0 = f_1 = 0$), an application of Green's 2nd identity to u and Δu gives

$$\int_D (u\Delta^2 u - (\Delta u)^2) = \int_\Gamma \left(u\frac{\partial\Delta u}{\partial\nu} - \frac{\partial u}{\partial\nu}\Delta u\right) ds = 0\,,$$

and hence $\Delta u = 0$ in D, and finally $u = 0$ by the uniqueness of the Dirichlet problem for the Laplace equation.

And we have shown that this pair of integral equations is satisfied, in particular, by $v = v_0$, $w = w_0$, where

$$v_0 = \Delta u \quad (\text{on } \Gamma)\,, \qquad w_0 = \frac{\partial \Delta u}{\partial \nu} \tag{*}$$

denote the Neumann data for the unique solution[2]) u to the biharmonic Dirichlet problem (I).

With any pair of continuous functions v, w on Γ we associate the constants

$$I_1 = I_1[v, w] = \int_\Gamma w \, ds\,,$$

$$I_3 = I_3[v, w] = \int_\Gamma \left(r^2 w - \frac{\partial(r^2)}{\partial \nu} v + 4f_1\right) ds\,, \qquad r(y) = |y|\,,$$

(of which the latter depends moreover on f_1 from (I)). If again u denotes the solution to (I) then the particular solution v_0, w_0 to (II) derived from u as in (*) satisfies $I_1 = I_3 = 0$:

$$I_1 = \int_\Gamma \frac{\partial \Delta u}{\partial \nu} ds = \int_D \Delta^2 u = 0\,,$$

$$I_3 = \int_\Gamma \left(r^2 \frac{\partial \Delta u}{\partial \nu} - \frac{\partial(r^2)}{\partial \nu} \Delta u + 4 \frac{\partial u}{\partial \nu}\right) ds = \int_D (r^2 \Delta^2 u - \Delta(r^2)\, \Delta u + 4\Delta u) = 0\,.$$

Our aim, however, is to study the pair of integral equations (II) independently, that is, we do not confine ourselves a priori to the solution (*) stemming from the solution to the Dirichlet problem (I) but consider the (affine space of) all solutions v, w to (II).

When speaking of a solution (v, w) to (II) it is tacitly understood that $v \in C^{1+\beta}(\Gamma)$ and $w \in C^\beta(\Gamma)$ for some $\beta > 0$.

Our main result is the following theorem concerning the pair of integral equations (II), confirming the conjecture in [2]:

Theorem: *Let d denote the logarithmic capacity (= outer mapping radius) of D* (cf. § 2). *Suppose that Γ and f_0 are of class $C^{4+\alpha}$ and f_1 of class $C^{3+\alpha}$ for some $\alpha > 0$, and denote by (v_0, w_0) the particular solution (*) to* (II) *stemming from the unique solution u to* (I).

a) *If $d \neq 1$* and *$d \neq e^{-1}$ then (v_0, w_0) is the only solution to* (II).

b) *Regardless of the value of d, (v_0, w_0) is the only solution to* (II) *such that $I_1 = I_3 = 0$.*

c) *If $d = 1$ or $d = e^{-1}$ then the solutions to* (II) *form a 1-dimensional affine space parametrized by I_1 and I_3, respectively.*

For the proof of this theorem it suffices to consider the homogeneous case

$$f_0 = f_1 = 0 \quad \text{on } \Gamma\,,$$

where $u = 0$ and hence $(v_0, w_0) = (0, 0)$.

The proof of a) will be reduced in § 4 and § 5 to the proof of b). We show, in fact, that for any solution (v, w) to (II), $d \neq 1$ implies $I_1 = 0$ (§ 4), and $d \neq e^{-1}$ plus $I_1 = 0$ imply $I_3 = 0$ (§ 5).

The proof of b) is given in § 6, and that of c) in § 7 and § 8. As to a) and b), it suffices to assume that Γ be of class $C^{1+\alpha}$ for some $\alpha > 0$, rather than $C^{4+\alpha}$.

In § 9 we obtain similar results for another natural choice of fundamental solution for $-\Delta^2$, viz. $G' = (8\pi)^{-1} \times \varrho^2(1 + \log(1/\varrho))$, $\varrho = |x - y|$. The main difference is that non-uniqueness for the solutions to the corresponding pair of integral equations occurs for $d = 1$, and for no other value of d.

In § 10 it is shown that, in dimensions $n > 2$, there is always a unique solution to the pair of integral equations associated in the above way with the biharmonic Dirichlet problem. The reason is that, in higher dimensions, the fundamental solution g for $-\Delta$ takes positive values only, while in the plane case, g is the logarithmic kernel, whose values are arbitrary real numbers. (The fact that the fundamental solution G to $-\Delta^2$ in 4 dimensions is logarithmic turns out to be of no consequence.)

2. The logarithmic capacity d and the equilibrium potential φ

It is well known that there exists a unique positive measure μ on Γ of total mass $\int_\Gamma d\mu = 1$ such that the logarithmic potential

$$\varphi(x) = \int_\Gamma \log \frac{1}{|x - y|} \, d\mu(y)$$

[2]) It is known that (I) has a (unique) solution $u \in C^{4+\alpha}(\Gamma)$ if Γ and f_0 are of class $C^{4+\alpha}$ and $f_1 \in C^{3+\alpha}(\Gamma)$ for some $\alpha > 0$, cf. [1, Theorem 12.12]. For the particular solution (v_0, w_0) in (*) we thus have $v_0 \in C^{2+\alpha}(\Gamma)$, $w_0 \in C^{1+\alpha}(\Gamma)$.

is constant (say equal to γ) on Γ (and hence also in $\complement D_e$ since φ is harmonic off Γ). This measure μ is called the *equilibrium measure* on Γ (or on $\bar{D}$). It is supported by Γ_e. The function φ is called the *equilibrium potential* for Γ (or $\bar{D}$). The positive number

$$d = e^{-\gamma}$$

is called the *logarithmic capacity*, or *outer mapping radius*, or *transfinite diameter* of Γ (or equally well of Γ_e, or of D or $\bar{D}$). Note that $\gamma - \varphi$ (> 0 in D_e) is Green's function for D_e with pole at ∞.

By application of the maximum principle it is seen that φ is characterized as the unique continuous function on $\mathbb{R}^2$ which is harmonic in $\complement\Gamma$, constant on Γ (hence in $\complement D_e$), and such that

$$\varphi(x) = \log \frac{1}{|x|} + O\left(\frac{1}{|x|}\right) \quad \text{as} \quad |x| \to \infty . \tag{1}$$

Alternatively, $\varphi = -\log |f|$, where $f\colon D_e \to \mathbb{R}^2$ ($= \mathbb{C}$) denotes the Riemann mapping function, carrying the simply connected domain $D_e \cup \{\infty\}$ onto $\{z \in \mathbb{C} \cup \{\infty\} \mid |z| < d\}$ and such that $f(\infty) = \infty$. Hence $|f'(\infty)| = 1$.

It is known that the equilibrium measure μ has a continuous density $q = d\mu/ds$ with respect to the arc length s on Γ, and that

$$q = -\frac{1}{2\pi}\left(\frac{\partial\varphi}{\partial\nu}\right)_+ , \tag{2}$$

where the plus means: from the outside E, and ν denotes the unit normal to Γ pointing away from D. See also the appendix, § 11.

For the convenience of the reader we bring here the simple proof that *the direct method of integral equations for the Dirichlet problem for the Laplace equation in the plane leads to non-uniqueness if and only if* $d = 1$. Green's 3rd identity tells us that

$$\int_\Gamma \left(g(x, y)\frac{\partial u(y)}{\partial\nu(y)} - \frac{\partial g(x, y)}{\partial\nu(y)} u(y)\right) ds(y) = \frac{1}{2} u(x)$$

for $x \in \Gamma$ when $u \in C^1(\bar{D})$ and u is harmonic in D. Writing $v = \partial u/\partial\nu$ and $f =$ the restriction of u to Γ, we arrive at the integral equation (of the first kind)

$$\int_\Gamma g(x, y)\, v(y)\, ds(y) = \frac{1}{2} f(x) + \int_\Gamma \frac{\partial g(x, y)}{\partial\nu(y)} f(y)\, ds(y) ,$$

serving to determine the Neumann data v from the Dirichlet data f on Γ.

In discussing uniqueness we may again confine ourselves to the homogeneous case $f = 0$. The integral equation then becomes

(i) $\quad \int_\Gamma g(x, y)\, v(y)\, ds(y) = 0 \quad \text{for} \quad x \in \Gamma .$

For any continuous function v on Γ the single layer potential

$$F(x) = \int_\Gamma g(x, y)\, v(y)\, ds(y)$$

is continuous in the whole plane (see § 11) and harmonic off Γ. Moreover, for large $|x| = R$, $-2\pi F(x) = I \log R + O(1/R)$ where $I = \int_\Gamma v\, ds$. Using (1) we find that $I\varphi - 2\pi F$ is $O(1/R)$, and of course continuous in $\mathbb{R}^2$, harmonic off Γ, and constant $= I\gamma$ on Γ (since $\varphi = \gamma$, $F = 0$ on Γ). From the maximum principle we infer that $I\varphi - 2\pi F \equiv I\gamma$ in $\mathbb{R}^2$, and since $I\varphi - 2\pi F$ is $O(1/R)$ we conclude that $I\gamma = 0$ and

$$F = \frac{1}{2\pi} I\varphi \quad \text{in the whole plane} .$$

If $d \neq 1$, i.e. $\gamma \neq 0$, this shows that $I = 0$ and hence $F = 0$ in $\mathbb{R}^2$. But that implies $v = 0$ on Γ since otherwise F would have a jump in its normal derivative when crossing Γ.

Thus we have found that the homogeneous integral equation (i) has $v = 0$ as its only continuous solution, when $d \neq 1$. And for any value of d, uniqueness is achieved by imposing the side condition $I = 0$, i.e. $\int v\, ds = 0$. Finally, for $d = 1$, the space of solutions is 1-dimensional. In fact, taking $v = q$ (the density of the measure μ generating φ), we get

$$F = \frac{1}{2\pi}\varphi ,$$

and so $v = q$ is the (unique) solution to the homogeneous integral equation with the side condition $I = 1$. — The proof in the biharmonic case is based on the same idea, but is considerably more involved, as we shall see.

3. The integrals F and f

The method under discussion is based on a choice of fundamental solutions $-G$ and $-g$ for the biharmonic operator Δ^2 and the Laplace operator Δ, respectively. Following [2], we choose

$$G = \frac{1}{8\pi} \varrho^2 \log \frac{1}{\varrho}, \qquad g = \frac{1}{2\pi} \log \frac{1}{\varrho},$$

where $\varrho = |x - y|$ is the distance between the generic points $x, y \in \mathbb{R}^2$. As mentioned in the introduction, G and g are related by

$$\Delta_x G(x, y) = \Delta_y G(x, y) = g(x, y) - \frac{1}{2\pi}. \tag{3}$$

With any pair (v, w) of continuous functions v, w on Γ we associate the following integral expressions $F = F[v, w] = F(x)$ and $f = f[v, w] = f(x)$, defining two functions F and f of the generic point $x \in \mathbb{R}^2$:

$$F = \int_\Gamma \left(Gw - \frac{\partial G}{\partial \nu} v\right) ds, \qquad f = \int_\Gamma \left(gw - \frac{\partial g}{\partial \nu} v\right) ds. \tag{4), (5}$$

Here $G = G(x, y)$, $\partial G/\partial \nu = \partial G(x, y)/\partial \nu(y)$, etc., and similarly for g, where $y \in \Gamma$ denotes the point of integration. From (3) follows

$$\Delta F = f - \frac{1}{2\pi} \int_\Gamma w \, ds, \qquad \Delta^2 F = \Delta f = 0 \quad \text{off } \Gamma. \tag{6}$$

It can be shown that F is of class C^1 in the whole plane (see § 11). It is well known that f is continuous relative to $\bar{D}$ and to $\bar{E}$. When crossing Γ from the inside D to the outside E, the single layer potential $\int gw ds$ remains continuous, while the double layer potential $\int (\partial g/\partial \nu) v ds$ jumps by $v(x)$, $x \in \Gamma$. Hence

$$f_+ - f_- = -v, \tag{7}$$

where the subscript $+$ or $-$ indicates approach to Γ from the outside E or the inside D, respectively.

For large $|x| = R$ we have the following expansions of F and f:

$$-8\pi F(x) = I_1 R^2 \log R + I_2(2R \log R + R) + I_3 \log R + I_4 + O(1/R), \tag{8}$$

$$-2\pi f(x) = I_1 \log R + I_2/R + (I_3 - I_4)/R^2 + O(1/R^3), \tag{9}$$

where (cf. [2])[3])

$$I_1 = \int_\Gamma w \, ds, \qquad I_2 = \int_\Gamma [(\xi\nu) v - (\xi y) w] \, ds, \qquad I_3 = \int_\Gamma [-2(\nu y) v + |y|^2 w] \, ds,$$

$$I_4 = \int_\Gamma \{-[2(\xi\nu)(\xi y) + (\nu y)] v + [\tfrac{1}{2}|y|^2 + (\xi y)^2] w\} \, ds.$$

Here y is the point of integration, $\nu = \nu(y)$ the outward unit normal, $\xi = x/|x|$ is the direction of x, and e.g., (ξy) denotes the inner product of the vectors ξ and y. Note that I_1 and I_3 are constants while I_2 and I_4 depend on ξ. Also note that I_1 and I_3 are the same as in § 1 (but now in the homogeneous case $f_0 = f_1 = 0$ on Γ).

The integral equations (II) now take the following form in the homogeneous case:

$$F[v, w] = 0, \qquad f[v, w] = \tfrac{1}{2} v \quad \text{on } \Gamma. \tag{10}$$

In §§ 4, 5, 6 below we suppose that Γ is of class $C^{1+\alpha}$ for some $\alpha > 0$, while in §§ 7, 8 we suppose Γ of class $C^{4+\alpha}$ for some $\alpha > 0$.

4. Consequences of (10) alone

Suppose that v, w are continuous functions on Γ satisfying the pair of integral equations (10), with F and f as defined in the whole plane by (4) and (5). We propose to establish (12), (13), and (14) below. According to (10) and (7),

$$f_+ = 0, \qquad f_- = v \quad \text{on } \Gamma. \tag{11}$$

For large $|x| = R$ we obtain from (9) and (1)

$$2\pi f = -I_1 \log R + O(1/R), \qquad I_1 \varphi = -I_1 \log R + O(1/R).$$

The difference $I_1\varphi - 2\pi f$ is thus $O(1/R)$ and moreover harmonic in E. Furthermore, $I_1\varphi - 2\pi f_+ = I_1\gamma - 0 = I_1\gamma$ on Γ according to § 2 and (11), and consequently $I_1\varphi - 2\pi f = I_1\gamma$ in E by the maximum principle. Since $I_1\varphi - 2\pi f = O(1/R)$, it follows that

$$I_1\gamma = 0,$$

[3]) Our I_4 is the I_5 from [2].

and consequently

$$f = \frac{1}{2\pi} I_1 \varphi \quad \text{in } E\,. \tag{12}$$

We have moreover established the implications:

If $d \neq 1$ (*i.e.* $\gamma \neq 0$) *then* $I_1 = 0$. *If* $I_1 = 0$ *then* $f = 0$ *in* E. (13), (14)

5. Consequences of (10) together with $I_1 = 0$

Suppose again that v, w are continuous and satisfy (10). We assume now further that $I_1 = 0$, that is, $\int w ds = 0$, and we propose to establish (17) and (18) below. According to (6) and (14), $\Delta F = f = 0$ in E, and hence, by (9),

$$I_2 = 0\,, \qquad I_4 = I_3\,. \tag{15}$$

Inserting this (and $I_1 = 0$) in (8) we obtain

$$-8\pi F = I_3(\log R + 1) + O(1/R)\,. \tag{16}$$

Comparing with (1) we infer (as in the proof of (12)) that $I_3\varphi - 8\pi F = I_3\gamma - 0 = I_3\gamma$ in E. On the other hand,

$$I_3\varphi - 8\pi F = I_3 + O(1/R)$$

by (1) and (16). It follows that

$$I_3 \cdot (1 - \gamma) = 0\,,$$

and consequently:

$$F = \frac{1}{8\pi} I_3 \cdot (\varphi - 1) \quad \textit{in } E\,. \tag{17}$$

If $d \neq e^{-1}$ (*i.e.* $\gamma \neq 1$) *and* $I_1 = 0$ *then* $I_3 = 0$. (18)

6. Uniqueness under the side conditions $I_1 = I_3 = 0$

Suppose now that $v \in C^{1+\alpha}(\Gamma)$ and $w \in C^\alpha(\Gamma)$ ($\alpha > 0$) satisfy the pair of integral equations (10) together with both side conditions $I_1 = I_3 = 0$. (Recall that, by (13) and (18), these side conditions are consequences of (10) if $d \neq 1$, e^{-1}.) First note that we now have

$$F = f = 0 \quad \text{in the whole plane}\,. \tag{19}$$

In E this follows from (17) and (12). To prove it in D, note that $\Delta^2 F = 0$ in D, by (6), and $F_- = (\partial F/\partial \nu)_- = 0$ on Γ because $F \in C^1(\mathbb{R}^2)$ and $F = 0$ in E. Moreover, $f \in C^1(\overline{D})$ because $v \in C^{1+\alpha}(\Gamma)$, $w \in C^\alpha(\Gamma)$, cf. [4, Satz 3 & 4, p. 99]. Since $\Delta F = f$ in D by (6), we thus have $\Delta F \in C^1(\overline{D})$ in the sense of Note 1, p. 446 above. Consequently, $F = 0$ in D by the uniqueness of the biharmonic Dirichlet problem, and $f = \Delta F = 0$ in D.

Next, it follows from (5) and (19) that

$$\int_\Gamma g w \, ds = \int_\Gamma \frac{\partial g}{\partial \nu} v \, ds \quad \text{in } \mathbb{R}^2\,.$$

Since the single layer potential $\int_\Gamma g w \, ds$ is continuous in $\mathbb{R}^2$, so is therefore the double layer potential $\int_\Gamma (\partial g/\partial \nu) v ds$. This implies that $v = 0$ (on Γ), and hence $w = 0$. — This completes the proof that (10) together with $I_1 = I_3 = 0$ imply $v = w = 0$.

7. Non-uniqueness in the case $d = e^{-1}$

In this section and the next we assume that Γ is of class $C^{4+\alpha}$ for some $\alpha > 0$, and we prove the existence of a solution (v, w) to (10) such that $I_3 = 1$ and (necessarily) $I_1 = 0$, cf. (13). This solution then spans the space of all solutions to (10) (according to § 6).

According to (17), the desired solution (v, w) must be such that

$$F = \frac{1}{8\pi}(\varphi - 1) \quad \text{in } \overline{E}\,, \tag{20}$$

and hence F is uniquely determined also in $\overline{D}$ as the solution to the biharmonic Dirichlet problem

$$\left.\begin{aligned} &\Delta^2 F = 0 \quad \text{in } D\,, \\ &F_- = 0 \quad \text{and} \quad \left(\frac{\partial F}{\partial \nu}\right)_- = \frac{1}{8\pi}\left(\frac{\partial \varphi}{\partial \nu}\right)_+ = -\frac{q}{4} \text{ on } \Gamma\,, \end{aligned}\right\} \tag{21}$$

where we have used (2), § 2. We therefore now define F by (20) in $\bar{E}$ and as the solution to (21) in $\bar{D}$. Since Γ is of class $C^{4+\alpha}$, so is φ on $\bar{E}$, and hence $q \in C^{3+\alpha}(\Gamma)$, see § 11. Consequently, $F \in C^{4+\alpha}(\bar{D})$, see [1, Theorem 12.12], and we may define functions $v \in C^{2+\alpha}(\bar{D})$ and $w \in C^{1+\alpha}(\Gamma)$ by

$$v = \Delta F \quad \text{in } \bar{D} = D \cup \Gamma, \qquad w = \partial v/\partial \nu \quad \text{on } \Gamma.$$

Since $\Delta v = \Delta^2 F = 0$ in D, we get from Green's 2nd identity together with (2):

$$I_3 = \int_\Gamma \left(-v \frac{\partial |y|^2}{\partial \nu} + |y|^2 \frac{\partial v}{\partial \nu}\right) ds = \int_D -v \Delta |y|^2 = -4 \int_D v = -4 \int_D \Delta F = -4 \int_\Gamma \frac{\partial F}{\partial \nu} ds$$

$$= \int_\Gamma q \, ds = \int_\Gamma d\mu = 1 .$$

In order to verify the latter integral equation (10) note that, since $\Delta v = 0$ in D,

$$\int_\Gamma \left(gw - \frac{\partial g}{\partial \nu} v\right) ds = \int_\Gamma \left(g \frac{\partial v}{\partial \nu} - \frac{\partial g}{\partial \nu} v\right) ds = \begin{cases} v \\ 0 \end{cases}$$

in D and E, respectively. On Γ the integral therefore equals $\frac{1}{2} v$.

As to the former equation (10) we obtain in E, using (3), (21), (2), and (20):

$$\int_\Gamma \left(Gw - \frac{\partial G}{\partial \nu} v\right) ds = \int_\Gamma \left(G \frac{\partial v}{\partial \nu} - \frac{\partial G}{\partial \nu} v\right) ds = -\int_D v \,\Delta G = -\int_D \left(g - \frac{1}{2\pi}\right) \Delta F =$$

$$= -\int_\Gamma \left[\left(g - \frac{1}{2\pi}\right) \frac{\partial F}{\partial \nu} - F \frac{\partial g}{\partial \nu}\right] ds = \int_\Gamma \left(g - \frac{1}{2\pi}\right) \frac{q}{4} ds = \frac{1}{8\pi} (\varphi - 1) = F \quad (\text{in } E) .$$

Approaching Γ from the outside E leads to the former integral equation (10), viz. $F[v, w] = 0$, since F vanishes continuously on Γ.

8. Non-uniqueness in the case $d = 1$

Supposing again that Γ is of class $C^{4+\alpha}$ for some $\alpha > 0$, we prove the existence of a solution (v, w) to (10) such that $I_1 = 1$. This time we omit the analysis of the problem (which is somewhat lengthy here) and head instead directly for the existence proof. The solution (v, w) is unique in view of (18) and § 6.

We shall need the *biharmonic potential* Φ of the equilibrium measure μ with density q (see § 2). It is defined for all $x \in \mathbb{R}^2$ by

$$\Phi(x) = \int_\Gamma G(x, y) \, d\mu(y) = \frac{1}{8\pi} \int_\Gamma |x - y|^2 \log \frac{1}{|x - y|} q(y) \, ds(y) . \tag{22}$$

Clearly Φ is continuous, and (first in the distribution sense)

$$2\pi \,\Delta \Phi = 2\pi \int_\Gamma \left(g - \frac{1}{2\pi}\right) d\mu = \varphi - 1 . \tag{23}$$

Actually, Φ is of class $C^{2+\alpha}$. Moreover, the restriction of Φ to $\bar{E}$ is of class $C^{4+\alpha}$.[4]) And so is therefore the unique bounded solution L to the Dirichlet problem in the unbounded set $\bar{E} = E \cup \Gamma$:

$$\Delta L = 0 \quad \text{in } E, \qquad L = \Phi \quad \text{on } \Gamma . \tag{24}$$

We now define a function F in the whole plane by[5])

$$F = \Phi - L + L(\infty)\,\varphi \quad \text{in } \bar{E} \tag{25}$$

(hence $F \in C^{4+\alpha}(\bar{E})$), and as the solution to the following biharmonic Dirichlet problem in $\bar{D}$:

$$\left.\begin{aligned} &\Delta^2 F = 0 \quad \text{in } D, \\ &F_- = 0 \quad \text{and} \quad \left(\frac{\partial F}{\partial \nu}\right)_- = \left(\frac{\partial F}{\partial \nu}\right)_+ \quad \text{on } \Gamma . \end{aligned}\right\} \tag{26}$$

---

[4]) Let B denote an open disk such that $B \supset \bar{D}$. According to (23), $-2\pi\Phi$ differs in B only by a harmonic function from the logarithmic potential $\psi(x) = \int_B g(x,y) \, (\varphi(y) - 1) dy$, because $\Delta\psi = -(\varphi - 1)$ in B. Since φ is Lipschitz continuous in $\bar{B}$, ψ is of class $C^{2+\alpha}$ in $\bar{B}$. This shows that $\Phi \in C^{2+\alpha}(\mathbb{R}^2)$. Since the restrictions of φ to $\bar{D}$ and $\bar{B} \cap \bar{E}$ are of class $C^{4+\alpha}$, those of ψ are of class $C^{5+\alpha}$, cf. [4, Satz 2, p. 102]. This shows that $\Phi \in C^{5+\alpha}(\bar{E})$ and $\Phi \in C^{5+\alpha}(\bar{D})$.

[5]) $L(\infty)$ denotes the limit value of L at infinity. By Green's 3rd identity (for $\bar{E}$):

$$L(\infty) = -\frac{1}{2\pi} \int_\Gamma L \frac{\partial \varphi}{\partial \nu} ds = -\frac{1}{2\pi} \int_\Gamma \Phi \frac{\partial \varphi}{\partial \nu} ds = \int_\Gamma \Phi \, d\mu .$$

Then $F \in C^1(\mathbb{R}^n)$, and the restriction of F to $\bar{D}$ is of class $C^{4+\alpha}$ according to [1, Theorem 12.12] because $(\partial F/\partial\nu)_+ \in C^{3+\alpha}(\Gamma)$. Finally put

$$v = \Delta F + \frac{1}{2\pi} \text{ in } \bar{D}, \qquad w = \frac{\partial v}{\partial \nu} + q \quad \text{on } \Gamma,$$

where q is given by (2). Clearly $v \in C^{2+\alpha}(\bar{D})$, $w \in C^{1+\alpha}(\Gamma)$.

Since v is harmonic in D, we have

$$I_1 = \int_\Gamma w \, ds = \int_\Gamma \frac{\partial v}{\partial \nu} ds + \int_\Gamma q \, ds = 0 + \int d\mu = 1 .$$

To verify the latter equation (10), note that

$$\int_\Gamma \left(gw - \frac{\partial g}{\partial \nu} v\right) ds = \int_\Gamma g \, d\mu + \int_\Gamma \left(g \frac{\partial v}{\partial \nu} - \frac{\partial g}{\partial \nu} v\right) ds$$

equals $(2\pi)^{-1}\varphi$ in E and $v + (2\pi)^{-1}\varphi$ in D. Since $\varphi = \gamma = 0$ on Γ, the left hand integral indeed equals $\frac{1}{2} v$ on Γ.

It remains to verify the former equation (10). In E we have in view of (22) and (3), writing $|y| = r$ for $y \in \bar{D}$:

$$\int_\Gamma \left(Gw - \frac{\partial G}{\partial \nu} v\right) ds = \int_\Gamma G \, d\mu + \int_D (G\Delta v - v\Delta G) = \Phi - \int_D \left(g - \frac{1}{2\pi}\right) v =$$

$$= \Phi - \int_D \left(g - \frac{1}{2\pi}\right) \Delta \left(F + \frac{r^2}{8\pi}\right) = \Phi - \int_\Gamma \left\{\left(g - \frac{1}{2\pi}\right) \frac{\partial}{\partial \nu}\left(F + \frac{r^2}{8\pi}\right) - \left(F + \frac{r^2}{8\pi}\right)\frac{\partial g}{\partial \nu}\right\} ds .$$

Inserting $F = 0$ on Γ and

$$\left(\frac{\partial F}{\partial \nu}\right)_- = \left(\frac{\partial F}{\partial \nu}\right)_+ = \frac{\partial \Phi}{\partial \nu} - \left(\frac{\partial L}{\partial \nu}\right)_+ + L(\infty)\left(\frac{\partial \varphi}{\partial \nu}\right)_+$$

by (26) and (25), we obtain

$$\int_\Gamma \left(Gw - \frac{\partial G}{\partial \nu} v\right) ds = \Phi + \int_\Gamma \frac{r^2}{8\pi}\frac{\partial g}{\partial \nu} ds - \int_\Gamma \left(g - \frac{1}{2\pi}\right)\left[\frac{\partial}{\partial \nu}\left(\Phi + \frac{r^2}{8\pi}\right) - \left(\frac{\partial L}{\partial \nu}\right)_+ + L(\infty)\left(\frac{\partial \varphi}{\partial \nu}\right)_+\right] ds . \quad (27)$$

Here we have (in E):

$$\int_\Gamma \left(g - \frac{1}{2\pi}\right)\frac{\partial}{\partial \nu}\left(\Phi + \frac{r^2}{8\pi}\right) ds = \int_\Gamma \left(\Phi + \frac{r^2}{8\pi}\right)\frac{\partial g}{\partial \nu} ds \quad (28)$$

because $\Phi + r^2/8\pi$ is harmonic in D by (23):

$$2\pi\Delta\left(\Phi + \frac{r^2}{8\pi}\right) = \varphi - 1 + \Delta\frac{r^2}{4} = \varphi = \gamma = 0 \quad \text{in } D .$$

Applying Green's 3rd identity to $L - L(\infty)$, we find in E:

$$-\int_\Gamma \left(g - \frac{1}{2\pi}\right)\left(\frac{\partial L}{\partial \nu}\right)_+ ds = -\int_\Gamma L \frac{\partial g}{\partial \nu} ds + L - L(\infty) \quad (29)$$

because $L - L(\infty)$ is harmonic in E with limit value 0 at ∞, and hence $\int_\Gamma (\partial L/\partial\nu)_+ \, ds = -\int_E \Delta L = 0$. Finally, we obtain from (2)

$$\int_\Gamma \left(g - \frac{1}{2\pi}\right)\left(\frac{\partial \varphi}{\partial \nu}\right)_+ ds = -2\pi \int_\Gamma \left(g - \frac{1}{2\pi}\right) d\mu = 1 - \varphi . \quad (30)$$

Multiplying (30) by $L(\infty)$, and adding (28) and (29), we obtain the following sum of the right hand members

$$L - L(\infty)\,\varphi + \int_\Gamma \frac{r^2}{8\pi}\frac{\partial g}{\partial \nu} ds$$

because $\Phi = L$ on Γ by (24). Consequently (27) reads

$$\int_\Gamma \left(Gw - \frac{\partial G}{\partial \nu} v\right) ds = \Phi - L + L(\infty)\,\varphi = F \quad \text{in } E$$

according to (25). Since F vanishes continuously on Γ, this establishes the former relation (10).

9. Use of the fundamental solution G'

In this section we replace G by the following fundamental solution for $-\Delta^2$

$$G'(x, y) = \frac{1}{8\pi}|x - y|^2\left(\log\frac{1}{|x - y|} + 1\right),$$

which has the advantage that (3), (6), and (8) are replaced by

$$\Delta G' = g\,, \tag{3'}$$

$$\Delta F' = f\,, \qquad \Delta^2 F' = \Delta f = 0 \quad \text{off } \Gamma\,, \tag{6'}$$

$$-8\pi F' = I_1(R^2 \log R - R^2) + I_2(2R \log R - R) + I_3 \log R + (I_4 - I_3) + O(1/R)\,, \tag{8'}$$

respectively, with the same coefficients I_1, I_2, I_3, I_4 as before.

We shall study the integral equations

$$F'[v, w] = 0\,, \qquad f[v, w] = \tfrac{1}{2} v \quad \text{on } \Gamma\,. \tag{10'}$$

There is no change in § 4, while § 5 is modified as follows: *If* $I_1 = 0$ *then* $I_2 = 0$, $I_4 = I_3$ (as before), *and hence* from (8')

$$-8\pi F' = I_3 \log R + O(1/R)\,. \tag{16'}$$

Combined with (1) this leads in the usual way to $I_3\gamma = 0$ and moreover

$$F' = \frac{1}{8\pi} I_3\varphi \quad \text{in } E\,. \tag{17'}$$

If $d \neq 1$ (i.e. $\gamma \neq 0$) we obtain $I_3 = 0$ (in addition to $I_1 = 0$).

Like in § 6 we establish uniqueness for solutions to (10') under the side conditions $I_1 = I_3 = 0$.

It remains to investigate the case $d = 1$. First we proceed as in § 7 (despite the different value of d), and we show that there exists a solution v, w (necessarily unique) to (10') such that $I_1 = 0$, $I_3 = 1$. In view of (17') we have to define F' by

$$F' = \frac{1}{8\pi}\varphi \quad \text{in } \overline{E}\,, \tag{20'}$$

and in complete analogy with (21) in $\overline{D}$. Proceeding as in § 7 we define $v = \Delta F'$ in $\overline{D}$ and $w = \partial v/\partial \nu$ on Γ, and we find that indeed $I_3 = 1$, and that (10') holds. (In the proof of the latter equation (10'), g enters in place of $g - 1/(2\pi)$ on account of (3')). Clearly,

$$I_1 = \int_\Gamma w \, ds = \int_\Gamma \frac{\partial v}{\partial \nu} \, ds = \int_D \Delta v = 0\,.$$

Finally, in place of § 8, it can be shown that *there exists a solution v, w to* (10') *with $I_1 \neq 0$ if and only if*

$$\iint G'(x, y) \, d\mu(x) \, d\mu(y) = 0\,,$$

but this condition is in general not fulfilled for a domain D of outer mapping radius $d = 1$ (in particular not for a circle of radius 1, since in that case $\iint G'(x, y) \, d\mu(x) \, d\mu(y) = 1/(8\pi)$).

Summing up, in the case $d = 1$ the space of solutions v, w to (10') has dimension 1 or conceivably (in exceptional cases) dimension 2.

10. The case of higher dimensions

For a bounded domain $D \subset \mathbb{R}^n$ ($n \geqq 3$) the direct method of integral equations applies just as for $n = 2$, but we shall see that there is always uniqueness without side conditions when $n \geqq 3$. A fundamental solution for $-\Delta$ is

$$g(x, y) = \frac{1}{a_n |x - y|^{n-2}}, \qquad a_n = (n - 2)\,\omega_n\,,$$

where ω_n denotes the surface area of the unit sphere in $\mathbb{R}^n$, i.e., $\omega_n = 2\pi^{n/2}/\Gamma(n/2)$. (For example, $a_3 = \omega_3 = 4\pi$.) A fundamental solution for $-\Delta^2$ is now

$$G(x, y) = -\frac{1}{A_n |x - y|^{n-4}}\,, \qquad A_n = 2(n - 2)(n - 4)\,\omega_n\,,$$

for $n \neq 4$; and $G(x, y) = \dfrac{1}{A_4}\log|x - y|$, $A_4 = 8\pi^2$, for $n = 4$. Thus (3) in § 3 is replaced now by $\Delta G = g$, and (6) by

$$\Delta F = f\,, \qquad \Delta^2 F = \Delta f = 0 \quad \text{off } \Gamma\,,$$

while (7) carries over. In place of (8), (9) we now have [6])

$$-A_nF = I_1R^{4-n} + (n-4)\,I_2R^{3-n} + (n-4)(I_3 + (n-2)I_4)R^{2-n} + O(R^{1-n}),$$
$$-a_nf = I_1R^{2-n} + (n-2)I_2R^{1-n} + (n-2)(I_3 + nI_4)R^{-n} + O(R^{-1-n}),$$

where

$$I_1 = \int_\Gamma w\,ds, \qquad I_2 = \int_\Gamma [-(\xi\nu)v + (\xi y)w]\,ds,$$
$$I_3 = \int_\Gamma [(y\nu)\,v - \tfrac{1}{2}|y|^2\,w]\,ds, \qquad I_4 = \int_\Gamma [-(\xi y)\,(\xi\nu)\,v + \tfrac{1}{2}(\xi y)^2\,w]\,ds,$$

with ds referring to surface measure on Γ.

Let $\varphi(x) = \int g(x, y)\,d\mu(y)\ (>0)$ denote the equilibrium potential of unit mass for $\overline{D}$, that is, the unique continuous function on $\mathbb{R}^n$ which is harmonic in $\complement\Gamma$, constant (say equal to $\gamma > 0$) on Γ, and such that, for large $|x| = R$:

$$\varphi(x) = R^{2-n} + O(R^{1-n}).$$

The functions $h = a_nf + I_1\varphi$ is harmonic and $O(R^{1-n})$ in E, and $h_+ = I_1\gamma$ on Γ by (7) and (10). It follows by the maximum principle that $h = I_1\gamma$ in E, and hence $I_1\gamma = 0$, $I_1 = 0$, $h = 0$ and $f = 0$ in E. Consequently, $I_2 = 0$, $I_3 + nI_4 = 0$. Inserting this in the expansion of F, we obtain

$$A_nF = -\frac{2(n-4)}{n}\,I_3R^{2-n} + O(R^{1-n}).$$

Arguing as above, now with $h = A_nF + \dfrac{2(n-4)}{n}\,I_3\varphi$, we find that $F = 0$ in E. This implies uniqueness (i.e. $v = w = 0$ on Γ) in the same way as in § 6.

11. Appendix

We shall verify some of the smoothness assertions in the text (for the plane case). General references: [1], [3], [4], [6].

When $\overline{D} \cup \Gamma$ is of class $C^{k+\alpha}$ for some $k = 1, 2, \ldots$ and $0 \leq \alpha < 1$, the solution to the Dirichlet problem for the Laplace equation in $\overline{D} = D \cup \Gamma$ is of class $C^{k+\alpha}(\overline{D})$ provided that the prescribed boundary values of the solution are of class $C^{k+\alpha}(\Gamma)$ [3, p. 106]. It follows easily that the equilibrium potential φ in § 2 is of class $C^{k+\alpha}(\overline{E})$, and hence the density $q = -(2\pi)^{-1}\,(\partial\varphi/\partial\nu)_+$ is of class $C^{k-1+\alpha}(\Gamma)$.[7])

In the sequel we always suppose $0 < \alpha < 1$. When $D \cup \Gamma$ is of class $C^{k+\alpha}$ ($k = 1, 2, \ldots$), any single layer potential

$$x \to \int g(x, y)\,w(y)\,ds(y)$$

with continuous density w on Γ is continuous (even Hölder continuous) in the whole plane $\mathbb{R}^2$ [4, p. 42].[8])

We may now show that $F[v, w] \in C^1(\mathbb{R}^2)$ for any continuous v, w on Γ. The former term $\int Gw\,ds$ in (4), § 3, is of class $C^1(\mathbb{R}^2)$ because G is of class $C^1(\mathbb{R}^2 \times \mathbb{R}^2)$. The latter term

$$\int_\Gamma \frac{\partial G}{\partial \nu}\,v\,ds = \sum_{j=1}^{2} \int \frac{\partial G(x, y)}{\partial y_j}\,\nu_j(y)\,v(y)\,ds(y)$$

is the sum of two integrals

$$H_j(x) = \int_\Gamma \frac{\partial G(x, y)}{\partial y_j}\,h_j(y)\,ds(y) = -\int \frac{\partial G(x, y)}{\partial x_j}\,h_j(y)\,ds(y),$$

with $h_j = \nu_j v$ continuous on Γ. Clearly H_j is continuous in $\mathbb{R}^2$, again because G is C^1. Moreover, for $i = 1, 2$:

$$\frac{\partial}{\partial x_i}\,H_j(x) = -\int \frac{\partial^2 G(x, y)}{\partial x_i\,\partial x_j}\,h_j(y)\,ds(y) \tag{31}$$

in the distribution sense. In other words,

$$8\pi\,\frac{\partial}{\partial x_i}\,H_j = \frac{\partial^2(|x|^2 \log |x|)}{\partial x_i\,\partial x_j} * \mu_j$$

where $*$ means convolution, and μ_j denotes the measure on Γ with density h_j, $d\mu_j = h_j\,ds$. Writing $|x| = r$, we have

$$\frac{\partial^2}{\partial x_j^2}\,(r^2 \log r) = 2 \log r + 1 + 2\,\frac{x_j^2}{r^2}, \quad j = 1, 2; \quad \frac{\partial^2}{\partial x_1\,\partial x_2}\,(r^2 \log r) = \frac{2x_1x_2}{r^2}.$$

Being the potential of a single layer with continuous density h_j, the term $(\log r) * \mu_j$ is continuous in $\mathbb{R}^2$ (as remarked above). Since $1 + 2x_j^2/r^2$ and $2x_1x_2/r^2$ are bounded, it follows easily by Lebesgue's theorem on bounded convergence that the convolutions

$$\left(1 + \frac{2x_j^2}{r^2}\right) * \mu_j \quad \text{and} \quad \frac{2x_1x_2}{r^2} * \mu_j$$

are everywhere defined and continuous in $\mathbb{R}^2$.

Having thus shown that the partial derivatives of H_j of order 1 in the distribution sense are continuous functions, we conclude that the continuous function H_j is indeed of class $C^1(\mathbb{R}^2)$, and (31) holds in the classical sense everywhere.

---

[6]) For $n = 4$, R^{4-n} should be replaced by $\log R$, while $n - 4$ should be replaced by -1 elsewhere.

[7]) One may for instance view the restriction of φ to $\overline{B} \cap \overline{D}_e$ (B an open disk $\supset \overline{D}$) as the solution to the Dirichlet problem in $\overline{B} \cap \overline{D}_e$ with boundary values γ (constant) on Γ_e and φ (analytic) on ∂B.

[8]) To pass from the 3-dimensional case studied in [4] to the plane case with the logarithmic kernel g, use the argument in [6, p. 172 ff.].

References

1 AGMON, S.; DOUGLIS, A.; NIRENBERG, L., Estimates near the boundary for solutions of elliptic partial differential equations satisfying general boundary conditions, Commun. Pure Appl. Math. **12** (1959), 623—727.

2 CHRISTIANSEN, S.; HOUGAARD, P, An investigation of a pair of integral equations for the biharmonic problem, J. Inst. Math., Appl. **22** (1978), 15—27.

3 GILBARG, D.; TRUDINGER, N. S., Elliptic Partial Differential Equations of Second Order, Grundlehren 224. Springer-Verlag, Berlin-Heidelberg-New York 1977.

4 GÜNTER, N. M., Die Potentialtheorie und ihre Anwendung auf Grundaufgaben der mathematischen Physik, Teubner 1957.

5 HAYES, J.; KELLNER, R., The eigenvalue problem for a pair of coupled integral equations arising in the numerical solution of Laplace's equation, SIAM J. Appl. Math. **20** (1972), 503—513.

6 KELLOGG, O. D., Foundations of Potential Theory, Grundlehren 31. Springer-Verlag, Berlin-Heidelberg-New York 1929, 1967.

Eingereicht: 22. 5. 1980

Anschrift: Prof. Dr. BENT FUGLEDE, Københavns Universitets Matematiske Institut, Universitetsparken 5, DK-2100 København, Danmark

KLEINE MITTEILUNGEN

ZAMM 61, 461 – 463 (1981)

O. KÜHLEWIND

Rang-1-Änderungen bei verallgemeinerten Inversen

1. Einleitung

Bekanntlich [7] existiert für jede komplexwertige $(m \times n)$-Matrix A die MOORE-PENROSE-Inverse (MPI) A^+, die als Lösung der vier PENROSE-Gleichungen

$$AXA = A\,, \qquad XAX = X\,, \tag{1), (2}$$

$$(AX)^* = AX\,, \qquad (XA)^* = XA\,, \tag{3), (4}$$

(* bedeutet konjugiert komplex und transponiert) eindeutig bestimmt ist. Alle Inversen, die wenigstens (1) erfüllen, werden unter dem Begriff „*verallgemeinerte Inversen*" zusammengefaßt. Falls A nichtsingulär ist, gilt $A^+ = A^{-1}$. Addiert man zu einer gegebenen Matrix $A \in C^{m \times n}$ (C^m bezeichnet den m-dimensionalen, komplexen Vektorraum, $C^{m \times n} = C^m \times C^n$) das dyadische Produkt $\sigma c d^*$, wobei $c \in C^m$, $d \in C^n$ und σ ein Skalar ist, entsteht die geänderte Matrix $M = (A + \sigma c d^*)$. Die verallgemeinerten Inversen von M sind der Gegenstand dieser Untersuchung.

In [9] wurde der Sonderfall, daß A und M nichtsingulär sind, ausführlich behandelt.

Für eine Reihe von verallgemeinerten Inversen von M sind aus der Literatur bereits Änderungsformeln bekannt. Eine vollständige Übersicht über Rang-1-Änderungsformeln für alle verallgemeinerten Inversen von M (8 Typen) wurde bisher nicht angegeben. Diese Arbeit versucht, diesen Mangel auszugleichen.

2. Bezeichnungen

Seien $A \in C^{m \times n}$, $X \in C^{n \times m}$, $c \in C^m$, $d \in C^n$ und σ ein komplexer Skalar. Dann gelten die folgenden *Bezeichnungen*:

$k = Xc \in C^n\,, \qquad h = d^*X \in C^m\,,$

$s = 1/\sigma + d^*Xc\,, \qquad \sigma \neq 0.,$

I Einheitsmatrix,

o Nullvektor,

$u = (I - AX)\,c = Ec \in C^m\,,$

$v = d^*(I - XA) = d^*F \in C^n\,,$

$\|\cdot\|$ Euklidische Vektornorm,

$r(\cdot)$ Rang einer Matrix,

$R(\cdot)$ Spaltenraum,

$|\cdot|$ Absolutbetrag.

Mit *(i, j, k)-Inverse von A* wird diejenige Matrix X bezeichnet, die die PENROSE-Gleichungen (i), (j) und (k) erfüllt; Symbol: $A^{i,j,k}$.

Insbesondere haben sich für A^1 bzw. $A^{1,2,3,4}$ die Bezeichnungen A^- bzw. A^+ durchgesetzt. Überstrichene Größen sind konjugiert komplex. Außerdem wird die bekannte Tatsache benutzt, daß sich die MPI eines Vektors $o \neq y \in C^m$ durch die Beziehung

$$y^+ = y^*/(y^*y) = y^*/\|y\|^2$$

berechnen läßt.

3. Änderungsformeln

Für eine übersichtliche Erfassung aller Möglichkeiten einer Rang-1-Änderung einer Matrix A um das dyadische Produkt $\sigma c d^*$ erweist sich die folgende *Fallunterscheidung* als geeignet [5]:

1. $c \notin R(A)\,, \quad d \notin R(A^*)\,; \qquad r(A + \sigma c d^*) = r(A) + 1\,,$

$$\left.\begin{array}{lll} 2.\ c \in R(A)\,, & d \notin R(A^*)\,, & s = 0; \\ 3.\ c \in R(A)\,, & d \text{ beliebig}\,, & s \neq 0; \\ 4.\ c \notin R(A)\,, & d \in R(A^*)\,, & s = 0; \\ 5.\ c \text{ beliebig}\,; & d \in R(A^*)\,, & s \neq 0; \end{array}\right\} \; r(A + \sigma c d^*) = r(A)\,,$$

6. $c \in R(A)\,, \quad d \in R(A^*)\,, \quad s = 0; \; r(A + \sigma c d^*) = r(A) - 1\,.$

Satz 1: *Sei $X = A^-$, dann gelten für die 6 Teilfälle die Beziehungen*

$$M^- = A^- - ku^+E - Fv^+h + sFv^+u^+E\,, \tag{5.1}$$

$$M^- = A^- - Fv^+h\,, \qquad M^- = A^- - \frac{1}{s}\,kh\,, \tag{5.2), (5.3}$$

$$M^- = A^- - ku^+E\,, \qquad M^- = A^- - \frac{1}{s}\,kh\,, \tag{5.4), (5.5}$$

$$M^- = A^-\,. \tag{5.6}$$

Beweis: Siehe [5] unter Beachtung von $s = 1/\sigma + d^*A^-c$, $\sigma \neq 0$.

Sei nun X eine (1,2)-Inverse von A. Dann überzeugt man sich leicht von der Richtigkeit der Zerlegungen

$$\begin{pmatrix} A + \sigma c d^* & c \\ o & -1/\sigma \end{pmatrix} = \begin{pmatrix} I & o \\ h & 1 \end{pmatrix} \begin{pmatrix} A & u \\ v & -s \end{pmatrix} \begin{pmatrix} I & k \\ o & 1 \end{pmatrix} \begin{pmatrix} I & o \\ \sigma d^* & 1 \end{pmatrix} \tag{6}$$

und

$$\begin{pmatrix} A + \sigma c d^* & o \\ d^* & -1/\sigma \end{pmatrix} = \begin{pmatrix} I & \sigma c \\ o & 1 \end{pmatrix} \begin{pmatrix} I & o \\ h & 1 \end{pmatrix} \begin{pmatrix} A & u \\ v & -s \end{pmatrix} \begin{pmatrix} I & k \\ o & 1 \end{pmatrix}. \tag{7}$$

Aus der Nichtsingularität der Scherungsmatrizen in (6) und (7) ergeben sich für die Klasse der (1,2)-Inversen die Beziehungen

$$\begin{pmatrix} A + \sigma c d^* & c \\ o & -1/\sigma \end{pmatrix}^{1,2} = \begin{pmatrix} I & -k \\ -\sigma d^* & \sigma s \end{pmatrix} \begin{pmatrix} A & u \\ v & -s \end{pmatrix}^{1,2} \begin{pmatrix} I & o \\ -h & 1 \end{pmatrix} \tag{8}$$

und

$$\begin{pmatrix} A + \sigma c d^* & o \\ d^* & -1/\sigma \end{pmatrix}^{1,2} = \begin{pmatrix} I & -k \\ o & 1 \end{pmatrix} \begin{pmatrix} A & u \\ v & -s \end{pmatrix}^{1,2} \begin{pmatrix} I & -\sigma c \\ -h & \sigma s \end{pmatrix}. \tag{9}$$

Durch zweimaliges Anwenden der Ränderungsformel für die (1,2)-Inversen [4] lassen sich bei Zurückgriff auf $X = A^{1,2}$ dann für die einzelnen Fälle (1,2)-Inverse der Matrix

$$\begin{pmatrix} A & u \\ o & -s \end{pmatrix}$$

berechnen. Mit (8) und (9) erhält man so Ausdrücke für

$$\begin{pmatrix} A + \sigma c d^* & c \\ o & -1/\sigma \end{pmatrix}^{1,2} \quad \text{und} \quad \begin{pmatrix} A + \sigma c d^* & o \\ d^* & -\sigma/1 \end{pmatrix}^{1,2}. \tag{10}$$

Aus (10) lassen sich leicht mittels Reduktionsformeln [6] (1,2)-Inverse von M bestimmen. Es ergibt sich der

Satz 2: *Sei $X = A^{1,2}$, dann gelten für die 6 Teilfälle die Beziehungen*

$$M^{1,2}_{i,j} = A^{1,2} - ku^+E - Fv^+h + sFv^+u^+E\,, \tag{11.1}$$

$$M^{1,2}_{i} = A^{1,2} - Fv^+h\,, \tag{11.2}$$

$$M^{1,2}_{i} = A^{1,2} - \frac{1}{P}\,(Fv^*h + \bar{s}kh)\,, \qquad P = \|v\|^2 + |s|^2, \tag{11.3}$$

$$M^{1,2}_{j} = A^{1,2} - ku^+E\,, \tag{11.4}$$

$$M^{1,2}_{j} = A^{1,2} - \frac{1}{Q}\,(ku^*E + \bar{s}kh)\,, \qquad Q = \|u\|^2 + |s|^2, \tag{11.5}$$

$$M^{1,2}_{i,j} = A^{1,2} + \sigma kh\,. \tag{11.6}$$

Bemerkung: Die Indizes i bzw. j an der Matrix $M^{1,2}$ vermerken neben dem Hinweis, daß es sich um spezielle (1,2)-Inverse von M handelt, die benutzte Zerlegung (6) bzw. (7). Die zweite Zahl der Gleichungsnummer benennt auch schon in (5) den Teilfall. Entsprechendes gilt nachfolgend für weitere verallgemeinerte Inverse von M.

Die Gleichungen (11.3) bzw. (11.5) wurden für $v = 0$ bzw. $u = 0$ in [6] und [8], S. 71, in Blockstruktur angegeben. Die Blockänderungsformeln von HARTWIG [2] für die (1,2)-Inversen beinhalten die Fälle 1, 2, 4 und 3 ∩ 5.

Satz 3: *Sei $X = A^{1,2,3}$, dann gelten für die 6 Teilfälle die Beziehungen*

$$M^{1,2,3}_{i,j} = A^{1,2,3} - ku^+ - Fv^+h + sFv^+u^+\,, \tag{12.1}$$

$$M^{1,2,3}_{i} = A^{1,2,3} - Fv^+h\,, \tag{12.2}$$

$$M_i^{1,2,3} = A^{1,2,3} - \frac{1}{P}(Fv^*h + \bar{s}kh)\,, \qquad P = ||v||^2 + |s|^2\,, \quad (12.3)$$

$$M_j^{1,2,3} = A^{1,2,3} - ku^+ - A^{1,2,3}h^+h\,, \quad (12.4)$$

$$M_j^{1,2,3} = A^{1,2,3} +$$

$$+ \frac{1}{Q}(sA^{1,2,3}h^*u^* - ||h||^2\,ku^* - ||u||^2\,A^{1,2,3}h^*h - \bar{s}kh)\,,$$

$$Q = |s|^2 + ||h||^2 \cdot ||u||^2\,, \quad (12.5)$$

$$M_i^{1,2,3} = A^{1,2,3} - A^{1,2,3}h^+h\,. \quad (12.6)$$

Beweis: Das Erfülltsein der dritten Penrose-Bedingung ergibt sich nach einiger Rechenarbeit unmittelbar aus den Gleichungen (13):

$$M \cdot M_{i,j}^{1,2,3} = AA^{1,2,3} + uu^+\,, \quad (13.1)$$

$$M \cdot M_i^{1,2,3} = AA^{1,2,3}, \quad (13.2)$$

$$M \cdot M_i^{1,2,3} = AA^{1,2,3}\,, \quad (13.3)$$

$$M \cdot M_j^{1,2,3} = AA^{1,2,3} - h^+h + uu^+\,, \quad (13.4)$$

$$M \cdot M_j^{1,2,3} = AA^{1,2,3} - h^+h + p_2^+p_2\,, \qquad p_2 = \frac{||h||^2}{s}\,u^* + h\,, \quad (13.5)$$

$$M \cdot M_i^{1,2,3} = AA^{1,2,3} - h^+h\,. \quad (13.6)$$

Mit (13) lassen sich dann auch (1) und (2) für die geänderte Matrix leicht verifizieren.

Satz 4: *Sei $X = A^{1,2,4}$, dann gelten für die 6 Teilfälle die Beziehungen*

$$M_{i,j}^{1,2,4} = A^{1,2,4} - ku^+E - v^+h + sv^+u^+E\,, \quad (14.1)$$

$$M_j^{1,2,4} = A^{1,2,4} - v^+h - kk^+A^{1,2,4}\,, \quad (14.2)$$

$$M_j^{1,2,4} = A^{1,2,4} +$$

$$+ \frac{1}{P}(sv^*k^*A^{1,2,4} - ||k||^2\,v^*h - ||v||^2\,kk^*A^{1,2,4} - \bar{s}kh)\,,$$

$$P = |s|^2 + ||k||^2 \cdot ||v||^2\,, \quad (14.3)$$

$$M_i^{1,2,4} = A^{1,2,4} - ku^+E\,, \quad (14.4)$$

$$M_i^{1,2,4} = A^{1,2,4} - \frac{1}{Q}(ku^*E + \bar{s}kh)\,, \qquad Q = ||u||^2 + |s|^2\,, \quad (14.5)$$

$$M_j^{1,2,4} = A^{1,2,4} - kk^+A^{1,2,4}\,. \quad (14.6)$$

Beweis: Entsprechend dem von Satz 3 unter Benutzung der Gleichungen (15):

$$M_{i,j}^{1,2,4}M = A^{1,2,4}A + v^+v\,, \quad (15.1)$$

$$M_j^{1,2,4}M = A^{1,2,4}A - kk^+ + v^+v\,, \quad (15.2)$$

$$M_j^{1,2,4}M = A^{1,2,4}A - kk^+ + p_1p_1^+, \qquad p_1 = \frac{||k||^2}{\bar{s}}\,v^* + k\,, \quad (15.3)$$

$$M_i^{1,2,4}\,M = A^{1,2,4}A\,, \quad (15.4)$$

$$M_i^{1,2,4}M = A^{1,2,4}A\,, \quad (15.5)$$

$$M_j^{1,2,4}M = A^{1,2,4}A - kk^+\,. \quad (15.6)$$

Bemerkung: Die Rang-1-Änderungsformeln (12) und (14) lassen sich unter Beachtung der Indizes ebenfalls aus (8) oder (9) unter Verwendung der Reduktionsformeln von Cline [1] nachweisen, falls man von $X = A^{1,2,3}$ bzw. $X = A^{1,2,4}$ ausgeht.

Folgerung: *Die Gleichungen* (12) *bzw.* (14) *behalten ihre Gültigkeit, falls* (auch in den Abkürzungen) (1,2,3) *durch* (1,3) *bzw.* (1,2,4) *durch* (1,4) *ersetzt wird.*

Beweis: Folgt unmittelbar, da sich (1) und (3) für M und $X = M^{1,2,3}$ bzw. (1) und (4) für M und $X = M^{1,2,4}$ ohne Verwendung von (2) für $X = A^{1,2,3}$ bzw. $X = A^{1,2,4}$ verifizieren lassen.

Die Gleichungen (12.3) für die (1,3)-Inversen bzw. (14.5) für die (1,4)-Inversen wurden für $u = 0$ bzw. $v = 0$ in [6] und [8], S. 71, in Blockstruktur angegeben.

Satz 5: *Sei $X = A^+$, dann gelten für die 6 Teilfälle die Beziehungen*

$$M^+ = A^+ - ku^+ - v^+h + sv^+u^+\,, \quad (16.1)$$

$$M^+ = A^+ - v^+h - kk^+A^+\,, \quad (16.2)$$

$$M^+ = A^+ + \frac{1}{P}(sv^*k^*A^+ - ||k||^2\,v^*h - ||v||^2\,kk^*A^+ - \bar{s}kh)\,, \quad (16.3)$$

$$P = |s|^2 + ||k||^2 \cdot ||v||^2\,,$$

$$M^+ = A^+ - ku^+ - A^+h^+h\,, \quad (16.4)$$

$$M^+ = A^+ + \frac{1}{Q}(sA^+h^*u^* - ||h||^2\,ku^* - ||u||^2\,A^+h^*h - \bar{s}kh)\,, \quad (16.5)$$

$$Q = |s|^2 + ||h||^2 \cdot ||u||^2\,,$$

$$M^+ = A^+ - kk^+A^+ - A^+h^+h + (k^+A^+h^+)\,kh\,. \quad (16.6)$$

Beweis: Siehe [5] unter Beachtung von $s = 1/\sigma + d^*A^+c$, $\sigma \neq 0$.

Folgerung: *Sei $X = A^{1,3,4}$, dann behalten die Gleichungen* (16) *ihre Gültigkeit, wenn M^+ durch $M^{1,3,4}$ und A^+ durch ersetzt werden.*

Beweis: Die Bedingungen (1), (3) und (4) für M und $X = M^{1,3,4}$ lassen sich ohne Verwendung von (2) für $X = A^+$ verifizieren. Aus den Beziehungen

$$M \cdot M^{1,3,4} = AA^{1,3,4}\,, \quad (17.2)$$

$$M \cdot M^{1,3,4} = AA^{1,3,4}\,, \quad (17.3)$$

$$M^{1,3,4}M = A^{1,3,4}A\,, \qquad M^{1,3,4}M = A^{1,3,4}A\,, \quad (17.4), (17.5)$$

$$M \cdot M^{1,3,4} = AA^{1,3,4} - h^+h\,, \qquad M^{1,3,4}M = A^{1,3,4}A - kk^+ \quad (17.6)$$

und den Gleichungen (13) bzw. (15) in abgeänderter Form für die (1,3)- bzw. (1,4)-Inversen von M ist der Nachweis von (3) und (4) für M und $X = M^{1,3,4}$ ersichtlich. Die Bedingung (1) ergibt sich im Anschluß ebenfalls durch Verifizieren. Damit sind für alle 8 Typen von verallgemeinerten Inversen Rang-1-Änderungsformeln gegeben.

Die aufgelisteten Rang-1-Änderungsformeln haben verschiedene Vorteile. Einerseits ist die Struktur der speziellen Änderungsformeln übersichtlich und wird durch (wenige) Fallunterscheidungen noch verstärkt. Andererseits treten gegenüber Blockänderungsformeln keine unbekannten (verallgemeinerten) Inversen auf, die neu zu berechnen wären. Das führt dazu, daß die Multiplikationszahlen sehr niedrig sind ($O(n^2)$).

Auch ist es ja eine bekannte Tatsache, daß sich jede Änderung vom Rang p durch eine Folge von p Änderungen vom Rang 1 ausdrücken läßt.

Neben ersten Anfängen [6], [8] stellt erst die Arbeit von Meyer [5] eine größere Zahl von Rang-1-Änderungsformeln bereit, die zusammen mit den Formeln von Hartwig [2], [3] im wesentlichen auch den derzeitigen Erkenntnisstand auf diesem Gebiet darstellen. Diese Arbeit versucht, noch bestehende Lücken auf dem Gebiet der Rang-1-Änderungsformeln fur verallgemeinerte inverse Matrizen zu schließen.

Zur allgemeinen historischen Entwicklung von verallgemeinerten Inversen werden zum Beispiel von Zielke [10] Ausführungen gemacht.

Der Verfasser bedankt sich an dieser Stelle bei Herrn Doz. Dr. G. Zielke, der durch eine Vielzahl von Hinweisen sowie durch Bereitstellung von Literatur am Entstehen dieser Arbeit unmittelbaren Anteil hat.

4. Literatur

1 Cline, R. E., Representations for the generalized inverse of a partitioned matrix, J. Soc. Indust. Appl. Math. 12 (1964), 588–600.
2 Hartwig, R. E., Block generalized inverses, Arch. Rat. Mech. Analysis 61 (1976), 197–251.
3 Hartwig, R. E., Perturbation theory. II: Similarities and differences between rank-one and row-column perturbations, Dept. of Math., North Carolina State Univ., Raleigh, North Carolina 1977.
4 Meyer, C. D., Jr., Representations for (1)- and (1,2)-inverses for partitioned matrices, Linear Algebra Appl. 4 (1971), 221–232.
5 Meyer, C. D., Jr., Generalized inversion of modified matrices, SIAM J. Appl. Math. 24 (1973), 315–323.
6 Mitra, S. K.; Bhimasankaram, P., Generalized inverses of partitioned matrices and recalculation of least squares estimates for data or model changes, Sankhyā, Ser. A 33 (1971), 395–410.
7 Penrose, R., A generalized inverse for matrices, Proc Cambridge Philos. Soc. 51 (1955), 406–413.

8 RAO, C. R.; MITRA, S. K., Generalized Inverse of Matrices and its Applications, John Wiley & Sons, New York 1971.
9 ZIELKE, G., Numerische Berechnung von benachbarten inversen Matrizen und linearen Gleichungssystemen, Vieweg & Sohn, Braunschweig 1970.
10 ZIELKE, G., Zur historischen Entwicklung von verallgemeinerten inversen Matrizen, Wiss. Z. Univ. Halle XXVII'78 M, H. 4, 109–118.

Eingereicht: 12. 5. 1980, revidierte Fassung: 3. 10. 1980

Anschrift des Autors: Dipl.-Math. OLAF KÜHLEWIND, Martin-Luther-Universität Halle-Wittenberg, Sektion Mathematik, Universitätsplatz 6, DDR-4020 Halle/Saale

ZAMM 61, 463–466 (1981)

K. UNSWORTH / T. C. CHIAM

A Numerical Solution of the Two-dimensional Boundary Layer Equations for Micropolar Fluids

1. Introduction

The inadequacy of the classical NAVIER-STOKES theory to describe rheologically complex fluids such as liquid crystals, polymeric fluids, animal blood, etc., has led to the development of microcontinuum fluid mechanics as an extension of the classical theory. Many models have been proposed to take into account the mechanically significant microstructure of such fluids. See [1, 2] for comprehensive reviews. One of the models, the theory of micropolar fluids introduced by ERINGEN [3], has generated a lot of interest in the literature. In this theory, the micropolar fluid exhibits the microrotational effects and microrotational inertia. Such fluids can support couple stress and body couples only. Physically they may represent fluids with bar-like or sphere-like elements.

Many flow problems involving micropolar fluids have been studied (see, for example, references [3, 4, 5, 6, 7, 8]) and the theory has also been applied to the study of blood flow [9, 10, 11]. Extension of the theory to include heat conduction has been made [12, 13], and applied to specific problems [12, 14, 15, 16].

In [17], WILLSON adapted the concept of a boundary-layer to flows of micropolar fluids past surfaces and obtained the appropriate two-dimensional boundary-layer equations, using an order-of-magnitude argument and neglecting certain microinertia terms. He then used the KÁRMÁN-POHLHAUSEN method to study the region near a stagnation point of a two-dimensional body. Using WILLSON's boundary-layer equations, NATH [18] obtained similarity solutions for flow past a wedge. However, as was pointed out by KÜMMERER [19], NATH's results are not correct and the only similarity solution possible for flow past a wedge is the case of stagnation point flow. In a subsequent paper [20], KÜMMERER used a RUNGE-KUTTA shooting method to solve a more general form of the similarity equations for stagnation point flow.

In this paper, we present solutions of the full boundary-layer equations for two-dimensional steady flow of a micropolar fluid past a wedge which, as far as the authors are aware, have not yet been studied in the available literature. Since analytical solutions to these equations are unlikely to be obtained, we have resorted to obtaining numerical solutions. The numerical method used is KELLER's 'box' method, a method which is known to be unconditionally stable, while also offering the possibility of producing extremely accurate solutions when used in conjunction with RICHARDSON extrapolation [21].

2. The Boundary-Layer Equations

The equations of motion for micropolar fluids have been derived by ERINGEN [3]. WILLSON [17] simplified these to obtain the equations of motion for steady two-dimensional incompressible boundary-layer flow in the absence of external body forces and body couples. Writing the velocity vector $\boldsymbol{v}$ and the microrotation vector $\boldsymbol{g}$ as

$$\boldsymbol{v} = [u(x, y), v(x, y), 0]\,, \qquad \boldsymbol{g} = [0, 0, g(x, y)] \tag{1}$$

where $Oxyz$ is a rectangular Cartesian coordinate system, y being measured outwards from the boundary into the main stream, and x along the boundary, he showed that the equations have the following form;

$$u\frac{\partial u}{\partial x} + v\frac{\partial u}{\partial y} = U_e\frac{dU_e}{dx} + \nu\frac{\partial^2 u}{\partial y^2} + K\frac{\partial g}{\partial y}\,, \tag{2}$$

$$G\frac{\partial^2 g}{\partial y^2} - 2g - \frac{\partial u}{\partial y} = 0\,, \qquad \frac{\partial u}{\partial x} + \frac{\partial v}{\partial y} = 0 \qquad (3), (4)$$

where U_e is the free-stream velocity, ν the apparent kinematic viscosity, K the coupling parameter and G the microrotation parameter. The boundary conditions to be satisfied by the unknowns are

$$\begin{aligned} &u(x, 0) = v(x, 0) = g(x, 0) = 0\,, \\ &u(x, y) \to U_e \quad \text{and} \quad g(x, y) \to 0 \quad \text{as} \quad y \to \infty \end{aligned} \tag{5}$$

3. Transformation of the Equations

NATH [18] proposed that similar solutions to equations (2)–(4), subject to the given boundary conditions (5), could be found if the free-stream velocity was of the form

$$U_e(x) = U_1 x^m \tag{6}$$

where U_1 and m are constants. KÜMMERER [19], however, demonstrated that this was only true for $m = 1$.

In order to study the development of the boundary layer in micropolar fluids, in which the free-stream velocity is given by (6), $m \neq 1$, we introduce the following transformations:

$$\psi(x, y) = \left(\frac{2\nu U_1}{m + 1}\right)^{1/2} x^{(m+1)/2} f_1(\eta, x)\,, \tag{7}$$

$$g(x, y) = \left(\frac{m + 1}{2\nu}\right)^{1/2} U_1^{3/2} x^{(3m-1)/2} f_2(\eta, x), \tag{8}$$

$$\eta(x, y) = y\left(\frac{m + 1}{2}\frac{U_1}{\nu}\right)^{1/2} x^{(m-1)/2} \tag{9}$$

where $\psi(x, y)$ is the stream function with $u = \frac{\partial\psi}{\partial y}$ and $v = -\frac{\partial\psi}{\partial x}$. Equations (2) and (3) now become

$$f_1''' + f_1 f_1'' + \beta(1 - f'^2) + K_1 f_2' = (2 - \beta)\, x\left[f_1'\frac{\partial f_1'}{\partial x} - f_1''\frac{\partial f_1}{\partial x}\right], \tag{10}$$

$$K_2 f_2'' - x^{1-m}(2f_2 + f_1'')(2 - \beta) = 0 \tag{11}$$

where the prime refers to differentiation with respect to η and

$$K_1 = \frac{K}{\nu}\,, \qquad K_2 = \frac{GU_1}{\nu}\,, \qquad \beta = \frac{2m}{m + 1}\,.$$

It would appear from equation (11), that for $m < 1$, the second term on the right-hand side would become increasingly dominant as $x \to \infty$. This, in turn, would indicate that $2f_2 + f_1'' \to 0$, i.e. $f_2 \to -\frac{1}{2}f_1''$, as $x \to \infty$. Results for small x, obtained in the early stages of the work, supported this observation; however it was clear that the final approach of $f_2 \to -\frac{1}{2}f_1''$ would only take place at very large values of x. Hence, in order to observe the behaviour of the solution for large x; it was decided to compress the range of x into a correspondingly smaller streamwise distance by introducing the variable ξ given by

$$\xi = \frac{1}{2}\ln(x + 1)\,.$$

Equations (10) and (11) now become:

$$f_1''' + f_1 f_1'' + \beta(1 - f_1'^2) + K_1 f_2' = \frac{1}{2}(2 - \beta)(1 - e^{-2\xi}) \times \left[f_1'\frac{\partial f_1'}{\partial \xi} - f_1''\frac{\partial f_1}{\partial \xi}\right], \tag{12}$$

$$K_2 f_2'' - (e^{2\xi} - 1)^{1-m}(2 - \beta)(2f_2 + f_1'') = 0\,. \tag{13}$$

The boundary conditions (5) may be rewritten as

$$f_1(0,\xi) = f_1'(0,\xi) = f_2(0,\xi) = 0\,. \qquad f_1'(\eta,\xi) \to 1 \quad \text{and} \quad f_2(\eta,\xi) \to 0 \quad \text{as} \quad \eta \to \infty \tag{14}$$

We now solve (12) and (13) subject to the boundary conditions (14).

4. The Numerical Method

The method used to solve (12) and (13) subject to (14), is fundamentally the same as the one used in [16] to solve a similar problem for heat conducting micropolar fluids. Briefly, (12) and (13) are each transformed into a system of first-order partial differential equations by the introduction of suitable new dependent variables. The systems are then put into finite difference form using KELLER's 'box' method [21]. In each case, the set of equations obtained can be put into vector-matrix form, in which the coefficient matrix has a block-tridiagonal structure. The two systems of equations, which are coupled, are then solved iteratively using a standard algorithm.

At a given ξ, a solution is initially found in the manner described above on a coarse mesh. The number of sub-intervals in the η-direction is then doubled by dividing each η sub-interval on the coarse mesh in half. A new solution is obtained on this finer mesh. Denoting two such solutions by S_{coarse} and S_{fine} say, a further solution, S, may be obtained using RICHARDSON extrapolation, viz;

$$S = \frac{1}{3}\,(4S_{\text{fine}} - S_{\text{coarse}}) \tag{15}$$

It is known that the solution (15) is fourth-order accurate in η and second-order accurate in ξ, whereas S_{coarse} and S_{fine} have only second-order accuracy in both variables.
The procedure now 'marches' to the next ξ-station, and the whole operation is repeated.

5. Results and Discussion

A wide range of results have been obtained using the method described in this paper. Some of the main results are summarised in Figures 1—7.

Figure 1 shows the development of the $f_1''(\eta,\xi)$ profile with downstream distance, for the case of flat-plate flow, i.e. $\beta = 0$, with $K_1 = 1.0$ and $K_2 = 10.0$. At $\xi = 0$, with $f_2(\eta, 0) \equiv 0$, the profile is that of similar flat-plate flow for a NEWTONian fluid. As the boundary-layer develops, however, the value of $f_1''(0,\xi)$ steadily increases and the value of η at which $f_1'(\eta,\xi) = 0$ decreases, so that at $\xi = 4.16$, where the program ceases calculation, there is a sharp turning point in $f_1'(\eta,\xi)$ quite close to

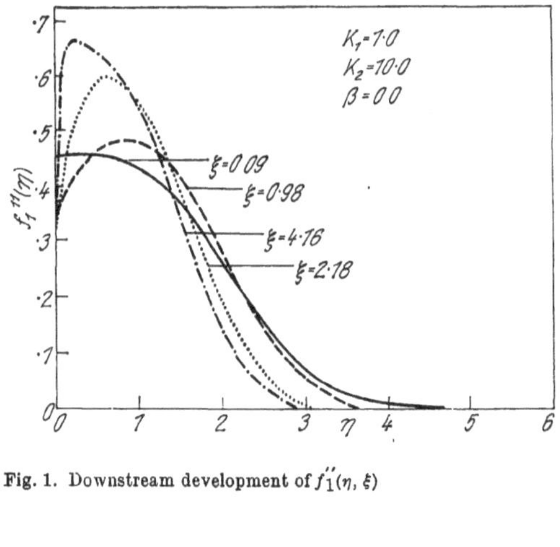

Fig. 1. Downstream development of $f_1''(\eta,\xi)$

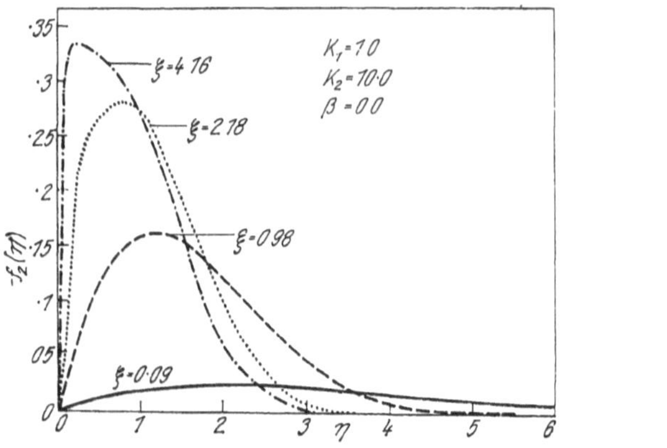

Fig. 2. Downstream development of $-f_2(\eta,\xi)$

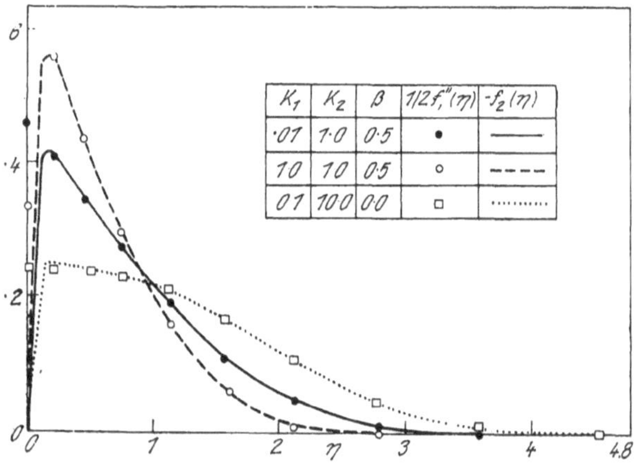

Fig. 3. Comparison of $f_1''(\eta,\xi)$ and $-f_2(\eta,\xi)$ at $\xi = 4.16$

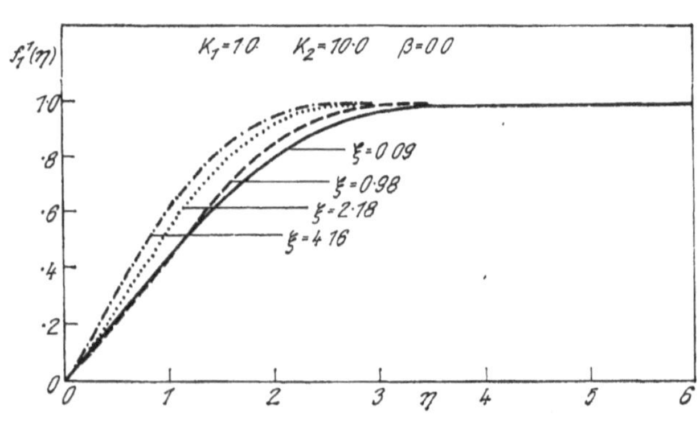

Fig. 4. Downstream development of velocity profile

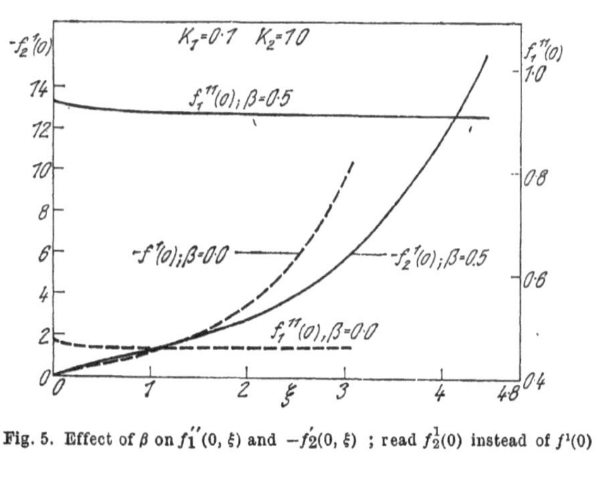

Fig. 5. Effect of β on $f_1''(0,\xi)$ and $-f_2'(0,\xi)$; read $f_2^1(0)$ instead of $f^1(0)$

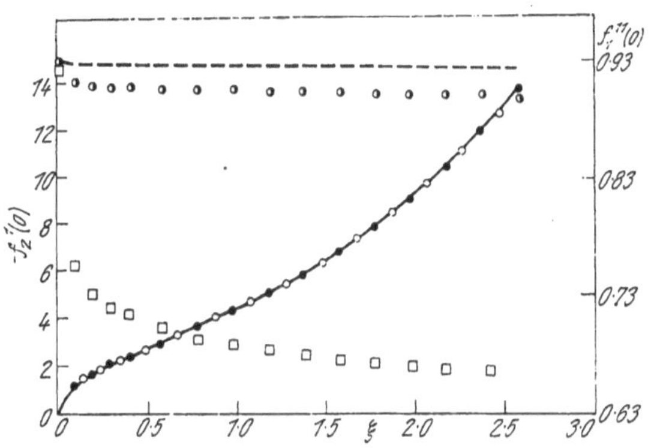

Fig. 6. Effect of K_1 on $f_1''(0, \xi)$ and $-f_2'(0, \xi)$ with $K_2 = 0.1$, $\beta = 0.5$. $f_2''(0, \xi)$; - - - - - -, ◑, □; $-f_2'(0, \xi)$; —, ● ○; for $K_1 = 0.01$, 0.1 and 1.0 respectively

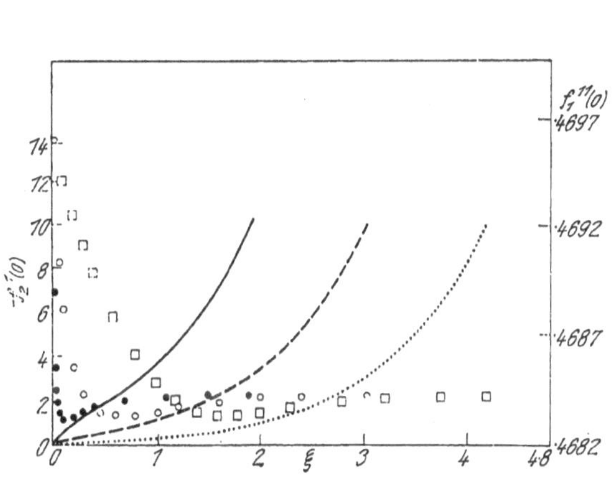

Fig. 7. Effect of K_2 on $f_1''(0, \xi)$ and $-f_2'(0, \xi)$ with $K_1 = 0.01$, $\beta = 0.0$ $f_1''(0, \xi)$; ●, ○, □; $-f_2'(0, \xi)$; ———, - - - - - -, ..., for $K_2 = 0.1, 1.0, 10.0$

the surface. Similar development may be found in Figure 2, where $-f_2(\eta, \xi)$ profiles are presented at different downstream locations. The shape of the $-f_2(\eta, \xi)$ profile at $\xi = 4.16$ can be seen to be very similar to that of $f_1''(\eta, \xi)$, with $[-f_2(\eta, \xi)]_{\max}$ about half of the value of $[f_1''(\eta, \xi)]_{\max}$. As noted earlier this behaviour is to be expected. As $\xi \to \infty$, for $m < 1$, the first term of (13) will become neglible, so that we then have

$$(e^{2\xi} - 1)^{1-m} (2f_2 + f_1'') (2 - \beta) = 0 . \qquad (16)$$

Clearly, (16) shows that $-f_2 \to \frac{1}{2} f_1''$ as $\xi \to \infty$. It should be noted, however, that $-f_2 \neq \frac{1}{2} f_1''$ at the boundary itself. Willson [17] studied this problem for small K_2 and conjectured the existence of a contact layer near the solid surface. Evidence supporting this theory is presented in Figure 3 where a comparison is made between the $-f_2(\eta, \xi)$ and $\frac{1}{2} f_1''(\eta, \xi)$ profiles at $\xi = 4.16$ for three different sets of values of β, K_1 and K_2. In each case, the large value of $-f_2'(0, \xi)$ is self-evident and there is very close agreement between the two profiles for most values of η. The existence of the contact layer is more obvious for the two profiles with $K_2 = 1.0$, the value of $-f_2'(0, \xi)$ in each case being significantly larger than that for $K_2 = 10.0$.

The downstream development of the velocity profile is shown in Figure 4, for flat-plate flow, with $K_1 = 1.0$ and $K_2 = 10.0$. The initial decrease in $f_1''(0, \xi)$ is offset by an increase of $f_1''(\eta, \xi)$ a short distance away from the surface, (see Figure 1), resulting in a corresponding decrease in the width of the boundary layer.

The remaining three figures attempt to isolate the effects of the parameters β, K_1 and K_2 on the development of the boundary-layer flow. In each case the variation of $-f_2'(0, \xi)$ and $f_1''(0, \xi)$ with ξ is shown for different values of the parameters, two of them kept constant.

Figure 5 demonstrates the effect of β for $K_1 = 0.1$ and $K_2 = 1.0$. Results are shown for $\beta = 0.0$ and 0.5. It can be seen that the value of $f_1''(0, \xi)$ stays practically constant in each case. If anything, there is a slight decrease in the value of $f_1''(0, \xi)$, accompanied by a small increase in $f_1''(\eta, \xi)$ a short distance out from the surface. The overall effect of β on the velocity profile is seen to be similar to that of this same parameter on flows of Newtonian fluids over wedges. The effect of β on $-f_2'(0, \xi)$ can be seen to be rather more significant; the expected increase in $-f_2'(0, \xi)$ is much more rapid for the smaller value of β, and the value of ξ at which the program stops is correspondingly less. A word should be said at this stage concerning the point at which the program ceases calculation. As the contact layer develops with distance downstream, a very large value of $-f_2'(0, \xi)$ appears, accompanied by a correspondingly large value of $-f_2'(\eta, \xi)$, with opposite sign, a short distance out from the surface. In order to calculate accurate detailed results in this region, a correspondingly finer mesh would be required. Our aim in this study is to obtain only a qualitative idea of the downstream development of the flow, so no such mesh refinement was undertaken. The calculation therefore reaches a stage, with the current mesh, at which unrealistic results are produced, and the program finally 'blows-up'.

A study of K_1 effects is presented in Figure 6. Results are shown for three different values of K_1, with $K_2 = 0.1$ and $\beta = 0.5$. The steady increase in $-f_2'(0, \xi)$ is again clear, but the effect of K_1 on this parameter is negligible. At the final ξ-station, the $-f_2(\eta, \xi)$ profiles are almost identical for $K_1 = 0.01$ and 0.1; but although $-f_2'(0, \xi)$ is almost the same at this station for $K_1 = 1.0$, the value of $[-f_2(\eta, \xi)]_{\max}$ is approximately 40% greater. The width of the boundary layer is also smaller in this case, thus providing clearer evidence of the existence of the contact layer. Starting at $\xi = 0$ with the similarity solution for Newtonian flow over a wedge, $f_1''(0, \xi)$ decreases in the early stages of the flow but then levels out, remaining virtually constant for $K_1 = 0.01$ and 0.1. At the final ξ-station, the shape of the $f_1''(\eta, \xi)$ profile has a significant dependence on the value of K_1. For $K_1 = 0.01$ and 0.1, $f_1''(\eta, \xi)$ is a monotonic decreasing function of η, whereas for $K_1 = 1.0$, it first of all increases, before falling away to zero. An idea of the different behaviour may be gathered from Figure 3.

Figure 7 demonstrates the effect of K_2 on the boundary layer development. The rate of increase in $-f_2'(0, \xi)$ is significantly faster for the smaller values of K_2, resulting in correspondingly smaller values of ξ at which the program stops. The qualitative behaviour of $f_1''(0, \xi)$ is the same in each case; an early decrease followed by a slight increase and then remaining almost constant. The changes in $f_1''(0, \xi)$ take place over a correspondingly smaller ξ-range for smaller values of K_2. The $f_1''(\eta, \xi)$ profiles at the final ξ-value in each case are virtually identical, implying very little K_2 influence on $f_1(\eta, \xi)$.

It should be mentioned at this point that the effects of K_1 and K_2 on flows with certain negative values of β are basically as described above. However, it is known, for Newtonian flow that a value of $\beta = -0.198838$ makes $f_1''(0, \xi) = 0$, leading to separation of the boundary layer. It is clear therefore that one effect of K_1 and K_2 will be to induce separation at negative values of β which have magnitude less than the value given above, the actual value of β depending on the choice of K_1 and K_2. This may be readily deduced from figure 6. However, a detailed investigation of this phenomenon is not included in the present work.

6. Conclusions

We have presented here results, obtained numerically, for the boundary-layer development of a micropolar fluid passing over an impermeable wedge. In all cases considered the development of the microrotation profile is qualitatively the same, starting with $f_2(\eta, 0) \equiv 0$, the wall gradient steadily increasing as the contact layer develops. The width of the velocity profile steadily decreases with distance downstream, a decrease in $f_1''(0, \xi)$ being accompanied by a corresponding increase in $f_1''(\eta, \xi)$ a short distance away from the wall. These changes in $f_1''(\eta, \xi)$ near or at the surface are significantly larger for larger values of K_1 and although K_1 has little effect on $-f_2'(0, \xi)$, the larger values of K_1 do produce more obvious evidence of the contact layer. K_2 has little effect on the velocity profile, but has a tendency to accelerate the development of the contact layer the smaller the value of K_2. The effect of β on the velocity profiles is seen to be the same as in NEWTONian flow, namely an increase in β brings an increase in $f_1''(0)$. The effect of β on the microrotation vector is similar to that of K_2.

Qualitatively, the results are in agreement with earlier work on the subject. The existence of the contact layer [17] is verified for the flows considered, while the effect of K_1 and K_2 on the velocity and microrotation vectors are in general agreement with the results of KÜMMERER [20]. It is hoped that results on heat conducting micropolar fluids passing over a wedge will be presented by the authors in the near future.

References

1 ARIMAN, T., TURK, M. A., and SYLVESTER, N. D., Int. J. Engr. Sci. **11**, 905 (1973).
2 COWIN, S. C., Adv. Appl. Mech. **14** C. S. YIH (ed.) 279 (Academic Press, New York, 1974).
3 ERINGEN, A. C., J. Math. Mech. **16**, 1 (1966).
4 WILLSON, A. J., J. Appl. Sci. Res. **20A**, 338 (1969).
5 ARIMAN, T. and CAKMAK, A. S., Rheol. Acta **7**, 236 (1968).
6 WILLSON, A. J., Proc. Camb. Phil. Soc. **64**, 513 (1968).
7 GUPTA, P. S. and GUPTA, A. S., Acta Mech. **15**, 141 (1972).
8 LIU, C. Y., Phys. Fluids **14**, 1808 (1971).
9 ARIMAN, T., J. Biomech. **4**, 185 (1971).
10 ARIMAN, T., TURK, M. A. and SYLVESTER, N. D., J. Appl. Mech. **41**, 1 (1974).
11 DEBNATH, L., Acta Mech. **24**, 165 (1976).
12 KAZAKIA, Y. and ARIMAN, T., Rheol. Acta **10**, 319 (1971).
13 ERINGEN, A. C., J. Math. Anal. and Appl. **38**, 480 (1972).
14 BALARAM, M. and SASTRY, V. U. K., Int. J. Heat Mass Transfer **16**, 437 (1973).
15 SASTRY, V. U. K. and MAITI, G., Int. J. Heat Mass Transfer **19**, 207 (1976).
16 UNSWORTH, K. and CHIAM, T. C., Rheol. Acta **19**, 356 (1980)
17 WILLSON, A. J., Proc. Camb. Phil. Soc. **67**, 469 (1970).
18 NATH, G., Rheol. Acta **14**, 850 (1975).
19 KÜMMERER, H., Rheol. Acta **15**, 584 (1976).
20 KÜMMERER, H., Rheol. Acta **16**, 261 (1977).
21 KELLER, H. B., Numerical Solutions of Partial Differential Equations — II B. HUBBARD (ed.) 327 (Academic Press, New York, 1971).

Anschrift: Dr. K. UNSWORTH, Dept. of Maths., Univ. of Dundee, Dundee DD1 4HN, Scotland, U. K.,
Dr. T. C. CHIAM, Dept. of Civ. Engg., Univ. College of Swansea, Swansea, Wales. U. K.

ZAMM 61, 466 – 469 (1981)

E. ERDOĞAN

Steady Pipe Flow of a Fluid of Fourth Grade

The steady flow of a fluid, which is called the *fourth-order fluid* or the *fluid of grade four*, in a circular pipe is considered. The velocity field of the flow is described by a first-order nonlinear differential equation and the solution of this differential equation is obtained in a closed form. The expressions of the flow rate and the apparent viscosity are given. It is found that the apparent viscosity decreases or increases depending on the material constants.

1. Introduction

Many materials such as drilling muds, clay coatings, and other suspensions, certain oils and greases, polimer melts, elastomers and many emulsions have been treated as non-NEWTONian fluids. Because of the difficulty to suggest a single model which exhibits all properties of non-NEWTONian fluids, they cannot be described as simply as NEWTONian fluid. Indeed, there has been much confusion over the classification of non-NEWTONian fluids. However, non-NEWTONian fluids may be classified as

(i) fluids for which the rate of shear depends only on the shear stress;

(ii) fluids for which the relation between shear stress and shear rate depends on time;

(iii) fluids which possess both elastic and viscous properties, the so-called visco-elastic fluids.

Because of the great diversity in the physical structure of non-NEWTONian fluids it is not possible to recommend a single constitutive equation as the equation for use in the cases described in (i), (ii) and (iii). For this reason, many non-NEWTONian models or constitutive equations have been proposed and most of them are empirical or semi-empirical. For more general three-dimensional representation, the constitutive equations of continuum mechanics are needed. Although many constitutive equations have been suggested many questions are still unsolved. Some of the continuum models do not give satisfactory results in accordance with the available experimental ones. For this reason, in many practical applications, empirical or semi-empirical equations have been used.

To understand easily usefulness of a constitutive equation it is usually natural that the circular pipe flow can be considered. Experimental results obtained in a circular pipe flow of a non-NEWTONian fluid show that the shape of the velocity profile near the axis becomes more flatter than that for a NEWTONian one. This property can be described by empirical or semi-empirical theories. Although the velocity field for a non-NEWTONian fluid in a circular pipe is different from that for a NEWTONian one, some continuum models cannot describe this property. For example, the fluid called the RIVLIN-ERICKSEN or second-order fluid, which is one of the simplest models of continuum mechanics, has the same velocity profile as that for a NEWTONian fluid in a circular pipe flow. However, the present paper shows that *the third order fluid and the higher ones have velocity profiles being different from that for a NEWTONian one.*

It has been shown by many authors that for many types of problems in which the flow is slow enough, in the visco-elastic sense, the results given by OLDROYD's constitutive equations will be substantially those of the second- or third-order RIVLIN-ERICKSEN constitutive equation. Thus, if this is the sense in which the solutions to which problems are to be interpreted, it would seem reasonable to use the second- or third-order RIVLIN-ERICKSEN equations in carrying out the calculations. This is particularly so in view of the fact that the calculations would generally be still simpler. For this reason, in the present paper, the fourth-order fluid is considered.

Laminar or turbulent flow of a non-NEWTONian fluid in a circular pipe has been investigated by many authors both to develop engineering techniques and applications and to improve mathematical and physical foundations, on the field of non-NEWTONian fluids. In this paper, the *steady laminar flow of a non-NEWTONian fluid in a circular pipe* is considered. A constitutive equation called the fourth-order fluid is used. It is a well-known fact that the second-order fluid equation for a pipe flow gives the same velocity profile as that for a NEWTONian one. However, experiment shows a different velocity profile with that for a NEWTONian one. Therefore, in this paper, the fourth-order fluid equation is used to investigate the flow in a circular pipe. It is shown that the third-order fluid equation gives the same velocity profile but a different pressure field compared with that for the fourth-order fluid equation. The fourth-order fluid equation has been used by many authors and a listing of some of the problems which have been solved may be found in [1].

Using a visco-elastic model the steady flow in a circular pipe has been investigated in [2]. The model in the present paper and that in [2] are not exactly the same. However, the differential equation satisfied by the velocity field has the same form both in the present paper and in [2], but one has to observe that the correspondence between the material coefficients is not complete. A closed form solution of this differential equation has not been given in [2], but a parametric solution has. Recently, a series solution in terms of the derivative of the stress normal to the cross-section has been given in [3]; however, terms of order four and higher orders in the expansion parameter are neglected. The present paper shows that *a series expansion solution cannot give all aspects of the problem considered.* For this reason, it seems to be reasonable to investigate for finding the closed form solutions of the governing equations. It is clear that results obtained in this paper involve those in [2], therefore a comparison between them is not needed.

In this paper, it is shown that the fourth-order fluid equation or the third-order one gives a non-linear differential equation for the velocity. This equation is a cubic one in terms of the derivative of the velocity and the solution, according to the author's knowledge, has not been given elsewhere. The solution can be established in a simple form. The distance to the axis of the pipe can be expressed easily in terms of the velocity component, but its inverse, namely the usual form, cannot be easily expressed; however, this is not the question. The velocity profiles depend on two parameters; one is always negative, the other may be positive or negative. The boundary condition for velocity at the wall gives a relation between these two parameters. A simple form of this relation can be obtained. For real materials, the values of these parameters are limited due to the expression of the velocity which is given in a square root form.

The investigation of the flow of a fluid in a circular pipe provides an opportunity to measure the material constants of the fluid model considered and a knowledge about the flow properties of the fluid. For this reason, using the expression of the flow rate, the apparent viscosity of the fluid is calculated. The apparent viscosity may be smaller or greater than that for the NEWTONian fluid depending on the material constants.

2. Basic equations

The fluid considered in this paper is a fourth-order fluid [4—6] and its *constitutive equation* is given by

$$\sigma = \sum_{i=1}^{4} \boldsymbol{S}_i \tag{1}$$

where

$$\boldsymbol{S}_1 = \mu \boldsymbol{A}_1 , \qquad \boldsymbol{S}_2 = \alpha_1 \boldsymbol{A}_2 + \alpha_2 \boldsymbol{A}_1^2 ,$$

$$\boldsymbol{S}_3 = \beta_1 \boldsymbol{A}_3 + \beta_2(\boldsymbol{A}_2\boldsymbol{A}_1 + \boldsymbol{A}_1\boldsymbol{A}_2) + \beta_3(\operatorname{tr} \boldsymbol{A}_2)\, \boldsymbol{A}_1 ,$$

$$\boldsymbol{S}_4 = \gamma_1 \boldsymbol{A}_4 + \gamma_2(\boldsymbol{A}_3\boldsymbol{A}_1 + \boldsymbol{A}_1\boldsymbol{A}_3) + \gamma_3 \boldsymbol{A}_2^2 + \gamma_4(\boldsymbol{A}_2\boldsymbol{A}_1^2 + \boldsymbol{A}_1^2\boldsymbol{A}_2) + \gamma_5(\operatorname{tr} \boldsymbol{A}_2)\, \boldsymbol{A}_2 + \gamma_6(\operatorname{tr} \boldsymbol{A}_2)\, \boldsymbol{A}_1^2 + [\gamma_7 \operatorname{tr} \boldsymbol{A}_3 + \gamma_8 \operatorname{tr}(\boldsymbol{A}_2 \boldsymbol{A}_1)]\, \boldsymbol{A}_1 ;$$

$\mu, \alpha_1, \alpha_2, \beta_1, \beta_2, \beta_3, \gamma_1, \gamma_2, \gamma_3, \gamma_4, \gamma_5, \gamma_6, \gamma_7, \gamma_8$ being material constants, $\boldsymbol{\sigma}$ the stress tensor, $\boldsymbol{I}$ the identity tensor and $\boldsymbol{A}_n$ representing the RIVLIN-ERICKSEN tensors defined by

$$\boldsymbol{A}_0 = \boldsymbol{I} , \qquad \boldsymbol{A}_1 = \nabla \boldsymbol{u} + (\nabla \boldsymbol{u})^{\mathrm{T}}$$

and

$$\boldsymbol{A}_{n+1} = \left(\frac{\partial}{\partial t} + \boldsymbol{u} \cdot \nabla\right) \boldsymbol{A}_n + (\nabla \boldsymbol{u}) \cdot \boldsymbol{A}_n + [(\nabla \boldsymbol{u}) \cdot \boldsymbol{A}_n]^{\mathrm{T}} .$$

In any viscometric flow there is $\boldsymbol{A}_3 = 0$ and $\boldsymbol{A}_4 = 0$ (cf., e.g, the discussion in [5]). Therefore, the material constants β_1, γ_1 $\gamma_2, \gamma_7, \gamma_8$ do not affect the velocity and the pressure fields in the flow considered in this paper. The *equation of motion*, in the absence of body forces, is

$$\varrho \frac{\mathrm{D}\boldsymbol{u}}{\mathrm{D}t} = -\nabla p + \nabla \cdot \boldsymbol{\sigma} \tag{2}$$

where ϱ is the density of the fluid, p the pressure, $\boldsymbol{u}$ the velocity and $\mathrm{D}/\mathrm{D}t$ the material derivative; and the *continuity equation for the velocity* is

$$\nabla \cdot \boldsymbol{u} = 0 . \tag{3}$$

The flow in a circular pipe is considered. Let the z-axis coincide with the axis of the pipe and cylindrical polar coordinates be used. The radius of the circular pipe is a. The flow in a circular pipe is generated under the action of a difference between the pressures at the two ends of the pipe. Thus, the *components of the velocity field* become

$$u_r = 0 , \qquad u_\theta = 0 , \qquad u_z = w(r)$$

where r, θ, z are cylindrical polar coordinates.

Substituting the velocity components in equation (1) gives

$$\sigma_{rr} = (2\alpha_1 + \alpha_2)\, w'^2 + 2[2(\gamma_3 + \gamma_4 + \gamma_5) + \gamma_6]\, w'^4 ,$$

$$\sigma_{\theta\theta} = 0 ,$$

$$\sigma_{zz} = \alpha_2 w'^2 + 2\gamma_6 w'^4 ,$$

$$\sigma_{rz} = \mu w' + 2(\beta_2 + \beta_3)\, w'^3 ,$$

$$\sigma_{r\theta} = 0 ,$$

$$\sigma_{\theta z} = 0 ,$$

where $\sigma_{rz} = \sigma_{zr}$, $\sigma_{r\theta} = \sigma_{\theta r}$, $\sigma_{\theta z} = \sigma_{z\theta}$, and a prime denotes differentiation with respect to r. Inserting the stress components and the velocity components in the equation (2), we obtain

$$\frac{\partial p}{\partial r} = (2\alpha_1 + \alpha_2) \frac{1}{r} (r w'^2)' + 2[2(\gamma_3 + \gamma_4 + \gamma_5) + \gamma_6)] \frac{1}{r} (r w'^4)' ,$$

$$\frac{\partial p}{\partial z} = \mu \left(w'' + \frac{1}{r} w'\right) + 2(\beta_2 + \beta_3) \left[(w'^3)' + \frac{1}{r} w'^3 \right] . \tag{4}$$

The continuity equation (3) is satisfied identically. From these two equations results that $\partial p/\partial z$ is constant. First the velocity field is obtained from equation (4) and then the pressure field is found. The equation (4) can be integrated once and the result may be written as

$$\mu w' + 2(\beta_2 + \beta_3)\, w'^3 = \frac{1}{2} r \frac{\partial p}{\partial z} + \frac{B}{r} , \tag{5}$$

where B is a constant. The equation (5) includes μ and $\beta_2 + \beta_3$ as material constants. Thus, there are no contributions of $\boldsymbol{S}_2$ and $\boldsymbol{S}_4$ to this equation. This means that *the velocity field of the third-order fluid and that for the fourth-order fluid are the same*; this statement is also correct for the first-order fluid and the second-order one.

The uniformity condition at $r = 0$ requires $B = 0$, and then the equation (5) becomes

$$\mu w' + 2(\beta_2 + \beta_3)\, w'^3 = \frac{1}{2} r \frac{\partial p}{\partial z} . \tag{6}$$

The boundary condition is

$$w(a) = 0$$

and $\partial p/\partial z$ is negative. The equation (6) may be written in non-dimensional form as

$$y' + \varepsilon y'^3 = \frac{1}{2} \beta x , \tag{7}$$

where

$$x = \frac{r}{a} , \qquad y = \frac{w}{W_0} , \qquad \varepsilon = \frac{2(\beta_2 + \beta_3)}{\mu (a/W_0)^2} ,$$

$$\beta = \frac{a\, \partial p/\partial z}{\mu W_0/a} , \tag{8}$$

and W_0 is the velocity at $r = 0$. The equation (7) has to satisfy the conditions

$$y(1) = 0 , \qquad y(0) = 1 , \tag{9}$$

where $0 \leqq x \leqq 1$. The equation (7) includes two parameters ε, β. β is negative and ε may be positive or negative. For $\varepsilon = 0$ the NEWTONian case is obtained. The solution of equation (7) is given in the next section.

3. The solution of the basic equation

The velocity field of the flow is given by equation (7) and it is subjected to the boundary condition (9). The differential equation (7) is a first-order non-linear one and it does not include y explicitly. Putting $s = y'$ we have

$$x = \frac{2}{\beta} (s + \varepsilon s^3) , \qquad y = \int s \, \mathrm{d}x + C , \tag{10}$$

where C is an arbitrary constant which will be given by condition (9). The equation (10) shows that $x = x(s)$ is a single-valued function; thus we have only one $x = 0$ for $s = 0$. Therefore, there is one horizontal tangent which is located at $x = 0$ and $y = 1$. Since β is negative, s is always negative in the interval considered. Since $y = 1$ for $s = 0$, then we have $C = 1$ and we write

$$y = 1 + \frac{1}{\beta} \left(s^2 + \frac{3}{2} \varepsilon s^4\right) . \tag{11}$$

The parametric solution of equation (7) is given by the equations (10) and (11). Although the values of x and y in terms of s can be given using the boundary conditions (9), the way which follows does not establish easily but it is straightforward. Therefore we try to express x in terms of y.

The equation (11) is a second-order algebraic equation in terms of s^2 and the solution can be given in the following form

$$s^2 = \frac{1}{3\varepsilon}\left\{[1 - 6\varepsilon\beta(1 - y)]^{1/2} - 1\right\};$$

where the other root is excluded, since $s = 0$ for $y = 1$. Further calculation gives

$$s = -\left\{\frac{[1 - 6\varepsilon\beta(1 - y)]^{1/2} - 1}{3\varepsilon}\right\}^{1/2},$$

since s is negative in the interval under consideration. Inserting the values of s^2 and s into the equation (10) we have

$$x = -\frac{2}{3\beta}\left(\frac{Y - 1}{3\varepsilon}\right)^{1/2}(Y + 2) \tag{12}$$

where

$$Y = [1 - 6\varepsilon\beta(1 - y)]^{1/2}.$$

The equation (12) gives the velocity field in a closed form. However, a relation between ε and β is needed. For $\beta = -4$, and $\varepsilon = 0$, we have $x = \sqrt{1 - y}$ which corresponds to the NEWTONian case. The other boundary condition which is $y(1) = 0$ gives

$$\beta = -\left[\frac{2}{3}\frac{(1 - 6\varepsilon\beta)^{1/2} - 1}{3\varepsilon}\right]^{1/2}\left[2 + (1 - 6\varepsilon\beta)^{1/2}\right]. \tag{13}$$

The equation (13) is a relation between ε and β. This relation will be given in a simpler form. Indeed, squaring both sides of the equation (13) and then after some algebra one finds

$$128\beta^2\varepsilon^2 + (27\beta^2 + 144\beta + 128)\,\beta\varepsilon + 8(\beta + 4) = 0.$$

This is a quadratic form in terms of ε and it has two roots; but only one root satisfies the condition $\varepsilon = 0$ for $\beta = -4$. Thus the result may be written in the following form

$$\varepsilon = \frac{1}{256\beta^2}\left\{[\beta(9\beta + 32)]^{3/2} - (27\beta^2 + 144\beta + 128)\,\beta\right\}. \tag{14}$$

The discussion on the righthandside of equation (14) can be summarized in Table 1; $\varepsilon \geqq 0$ for $\beta \leqq -4$ and $\varepsilon \leqq 0$ for $-4 \leqq \beta \leqq -32/9$.

When $\beta_c = -32/9$ and $\varepsilon_c = -3/64$ we have $6\beta_c\varepsilon_c = 1$ and the solution can be written as

$$x = \frac{1}{2}(4 - 3y - y^{3/2})^{1/2}. \tag{15}$$

The values ε_c and β_c correspond to the lower limit for ε and the upper limit for β, respectively; they may be called the *critical values* of ε and β. The corresponding differential equation for the solution given by (15) is

Table 1

β		$-(24 + 8\sqrt{3})/9$		-4		$-32/9$		$-(24 - 8\sqrt{3})/9$	
$27\beta^2 + 144\beta + 128$	+		−		−		−		+
β	−		−		−		−		−
$27\beta^2 + 144\beta + 128)\,\beta$	−		+		+		+		−
	$\varepsilon > 0$		$\varepsilon \geqq 0$		$\varepsilon \leqq 0$		$\varepsilon < 0$		$\varepsilon > 0$

$$y' - \frac{3}{64}y'^3 = -\frac{16}{9}x. \tag{16}$$

When ε tends to positive infinity β does to minus infinity, but their ratio becomes

$$\lim_{\varepsilon \to +\infty} \frac{\varepsilon}{\beta} = -\frac{27}{128},$$

and the corresponding solution is

$$\lim_{\varepsilon \to +\infty} x = (1 - y)^{3/4}$$

or $y = 1 - x^{4/3}$.

Table 2 shows the values of ε and β calculated from equation (14). The values given in Table 2 are plotted in figure 1. Small values of ε correspond to small values of $\beta + 4$. A series expansion of ε in terms of β can be given in the following form

Table 2

β	ε	β	ε	
β_c	ε_c	−4.71	0.100695	$\beta_c = -3.555555$
−3.60	−0.043394	−9.56	0 999967	$\varepsilon_c = -0.046875$
−3.70	−0.034120	−20	3.144545	
−3.90	−0.012121	−100	19.978780	
−4.00	0.000000	−1000	209.8135	$\lim_{\varepsilon \to +\infty} \frac{\varepsilon}{\beta} = -0.210937$

$$\varepsilon = -\frac{1}{8}(\beta + 4) + \frac{1}{32}(\beta + 4)^2 + \frac{1}{64}(\beta + 4)^3 + \dots \tag{17}$$

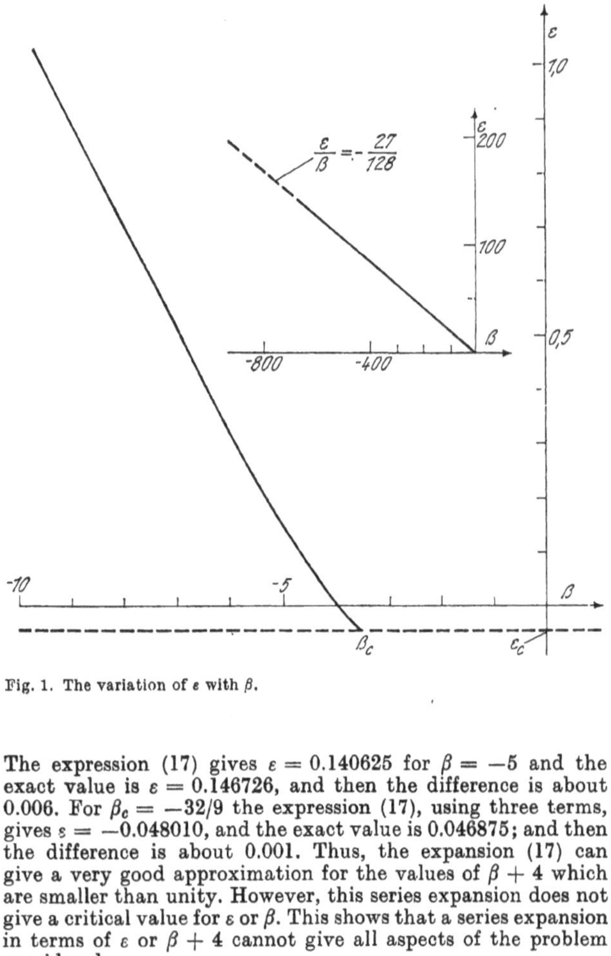

Fig. 1. The variation of ε with β.

The expression (17) gives $\varepsilon = 0.140625$ for $\beta = -5$ and the exact value is $\varepsilon = 0.146726$, and then the difference is about 0.006. For $\beta_c = -32/9$ the expression (17), using three terms, gives $\varepsilon = -0.048010$, and the exact value is 0.046875; and then the difference is about 0.001. Thus, the expansion (17) can give a very good approximation for the values of $\beta + 4$ which are smaller than unity. However, this series expansion does not give a critical value for ε or β. This shows that a series expansion in terms of ε or $\beta + 4$ cannot give all aspects of the problem considered.

The large values of ε correspond to the large values of $|\beta|$ and the relation between them can approximately be written as

$$\frac{\varepsilon}{\beta} = -\frac{27}{128} - \frac{9}{8\beta} + O(\beta^{-3}). \tag{18}$$

For $\beta = -50$ the expression (18) gives $\varepsilon = 9.422$ and the exact value is $\varepsilon = 9.442$; then the difference is about 0.02. For $\beta = -10$ from the equation (18) we have 0.984 and the exact value is about 1.088; then the difference is about 0.104. Thus, the expression (18) can give a very good approximation for the values of β when $\beta \leqq -10$.

Since ε is known in terms of β, x which is given in (12), can be expressed in terms of y. The corresponding values of x and y are given in Table 3. The values given in Table 3 are plotted in Figure 2.

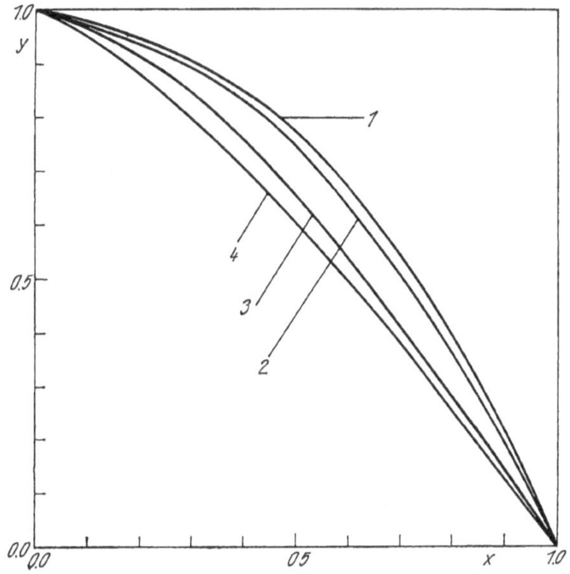

Fig. 2. The velocity profiles. ①: $x = \frac{1}{2}(4 - 3y - y^{3/2})^{1/2}$ for $\beta = -32/9$ and $\varepsilon = -3/64$; ②: $x = (1 - y)^{1/2}$ for $\beta = -4$ and $\varepsilon = 0$; ③: $x = 0.04\,\{2 + [1 + 57 \cdot 36(1 - y)]^{1/2}\}\,\{[1 + 57 \cdot 36(1 - y)]^{1/2} - 1\}^{1/2}$ for $\beta = -9.56$ and $\varepsilon = 1$; ④: $x = (1 - y)^{3/4}$ for $\beta = -\infty$ and $\varepsilon = +\infty$

Tabelle 3

	①	②	③	④
y	$\beta = -32/9$, $\varepsilon = -3/64$	$\beta = -4$, $\varepsilon = 0$	$\beta = -9.56$, $\varepsilon = 1$	$\beta = -\infty$, $\varepsilon = +\infty$
1.0	0.0000	0.0000	0.0000	0.0000
0.8	0.4702	0.4472	0.3543	0.2991
0.6	0.6586	0.6325	0.5476	0.5030
0.4	0.7980	0.7746	0.7124	0.6817
0.2	0.9097	0 8944	0.8614	0.8459
0.0	1.0000	1.0000	1.0000	1.0000

The curve ② corresponds to the velocity profile for a NEWTONian fluid flow. The curve ① shows the critical case. The values of x in the curve ① are greater than those in the curve ②. The fluid particles which correspond to the curve ① move easier than particles of a NEWTONian fluid under the same pressure gradient. This is not true for fluid particles which correspond to the curves ③ and ④. For small values of ε the curves can be plotted using a series expansion in the following form

$$y = 1 - x^2 - 2\varepsilon x^2(1 - x^2) + 4\varepsilon^2 x^2(1 - x^2)(1 + 4x^2) - 32\varepsilon^3 x^2(1 - x^2)(1 + x^2 + 6x^4) + \dots$$

or

$$y = 1 - x^2 + \frac{\beta + 4}{4} x^2(1 - x^2) + \frac{(\beta + 4)^2}{4} x^4(1 - x^2) - \frac{(\beta + 4)^3}{16} x^4(1 - x^2)(1 - 6x^2) + \dots,$$

since the small values of ε correspond to those of $|\beta + 4|$. For the large values of ε we have

$$y = 1 - x^{4/3} + \frac{3}{8\varepsilon}\left(x^{2/3} - x^{4/3}\right) - \frac{9}{64\varepsilon^2}\left(x^{2/3} - x^{4/3}\right) + O(\varepsilon^{-3})$$

or

$$y = 1 - x^{4/3} - \frac{16}{9\beta}\left(x^{2/3} - x^{4/3}\right) + \frac{512}{81\beta^2}\left(x^{2/3} - x^{4/3}\right) + O(\beta^{-3}).$$

The correspondence between the approximate values obtained from the above expressions and the exact ones is very good.

4. The flow rate and the apparent viscosity

It is a well known fact that the flow in a circular pipe can be used to measure the material constants of a fluid model. For this reason, it is necessary to calculate the flow rate and it is given by

$$\frac{Q}{2\pi a^2 W_0} = \int_0^1 yx \, dx,$$

where x and y are non-dimensional variables which are given by equation (8). Substituting x and dx in terms of y and dy from equation (12) into the flow rate expression we have

$$\frac{2Q}{\pi a^2 W_0} = -\frac{8}{405\varepsilon^2\beta^3}\left[(1 - 6\varepsilon\beta)^{5/2} + 135\varepsilon^2\beta^2 + 15\varepsilon\beta - 1\right] \quad (19)$$

where W_0 is the velocity on the axis of the pipe. Table 4 shows the values of the flow rate which correspond to some values of ε. The flow rate is non-dimensionalized by the flow rate for the case of the NEWTONian fluid. For this reason, the dimensionless flux equals to unity for $\varepsilon = 0$ and $\beta = -4$. For negative values of ε the flow rate is greater than that for the NEWTONian case. However, for positive values of ε it is smaller than that for the NEWTONian case. The dimensionless flow rate for the model considered in this paper varies between 0.8 and 1.05.

Table 4

	$\varepsilon = \varepsilon_c$, $\beta = \beta_c$	$\varepsilon = 0$, $\beta = -4$	$\varepsilon/\beta = -27/128$
$\frac{2Q}{\pi a^2 W_0}$	$\frac{21}{20}$	1	$\frac{4}{5}$

The apparent viscosity of the fluid in a circular pipe is defined as

$$Q = \frac{\left(\frac{-\partial p}{\partial z}\right)\pi a^4}{8\mu^*},$$

where μ^* is the apparent viscosity. Eliminating $\partial p/\partial z$ between Q and $\beta = a^2(\partial p/\partial z)/\mu W_0$ we obtain

$$\frac{\mu^*}{\mu} = \frac{-\beta/4}{2Q/\pi a^2 W_0}. \quad (20)$$

The values of the apparent viscosity for values of ε or β are given in Table 5. It can be seen from Table 5 that the apparent viscosity has the smallest value for $\varepsilon = \varepsilon_c$, and for the positive values of ε it has greater values than that for the NEWTONian case.

Table 5

β	ε	$2Q/\pi a^2 W_0$	μ^*/μ
$-\frac{32}{9}$	$-\frac{3}{64}$	$\frac{21}{20}$	$\frac{160}{189}$
-4	0	1	1
-10	1.1501	0.8232	3.0369
$\frac{\varepsilon}{\beta} = -\frac{27}{128}$		$\frac{4}{5}$	$\frac{\mu^*/\mu}{-\beta} = \frac{5}{16}$

References

1 BIRD, R. B., Ann. Rev. Fluid Mech. 8, 13 (1976).
2 SHERMA, S. K., Z. Angew. Math. Mech (ZAMM) 56, 313 (1959).
3 DUNWOODY, N. T.; HAMIL, T. A., Z. Angew Math. Phys. (ZAMP) 30, 587 (1979).
4 JOSEPH, D. D.; FOSDICK, R. L., Arch. Rat. Mech. Analysis 49, 321 (1973).
5 TRUESDELL, C., Ann. Rev. Fluid Mech. 6, 111 (1974).
6 BEAVERS, G. S.; JOSEPH, D. D., J. Fluid Mech. 69, 475 (1975).

Eingereicht: 8. 12. 1980

Address: Prof. Dr. M. EMIN ERDOĞAN, Istanbul Teknik Universitesi, Makina Fakültesi, Gümüşsuyu, Istanbul, Turkey

BUCHBESPRECHUNGEN

Lippmann, H., Extremum and Variational Principles in Mechanics, Course held at the Department of General Mechanics 1970, Wien-New York. Springer-Verlag. 1970. 239 S., 39 Abb., DM 39,10. S 270,–. (CISM Courses and Lectures 54)

Besdo, D., Example to Extremum and Variational Principles in Mechanics, Seminar Notes Accompaning the Vol. 54 by H. Lippmann, Wien-New York. Springer-Verlag. 1973. 236 S., 59 Abb., DM 40,30. S. 278,–. (CISM Courses and Lecture 65)

Hervorgegangen aus Vorlesungen am Internationalen Zentrum für mechanische Wissenschaften (CISM) vermitteln die Bücher die Grundideen der Variationsrechnung und Variationsprinzipe der Mechanik. Es ist das Anliegen der Verf., alle Begriffe und Aussagen an einfachen Anwendungen zu illustrieren. Zu diesem Zweck sind in nahezu allen Kapiteln Beispiele eingefügt. Diesem Ziel dient auch der Begleitband von D. Besdo, CISM-Bericht Nr. 65, der als Ergänzung betrachtet werden muß. Die theoretische Darstellung des Stoffes benutzt Mathematikkenntnisse, die im Rahmen der Ingenieurausbildung vermittelt werden. Neben einer Einführung über stationäre Punkte von Vektorfunktionen und ihre mathematische Beschreibung mit Hilfe des Variationskalküls werden folgende Gebiete untersucht:
Im Kap. „Allgemeine Dynamik" sind für diskrete Systeme das d'Alembertsche Prinzip, die Lagrangeschen Gleichungen 1. Art, die Begriffe Arbeit, Energie, Wirkung, die Prinzipien von Hamilton, Leibniz – Maupertuis und Jacobi beschrieben und die vorhandenen Zusammenhänge eingehend untersucht. Die entsprechenden Begriffsbildungen für Kontinuumsprobleme werden aus o. g. Ergebnissen durch Grenzübergang abgeleitet. Das Kap. „Variationsrechnung und Anwendungen auf die allgemeine Dynamik" enthält zunächst die notwendigen Bedingungen für die stationären Werte eines Funktionals und die daraus entstehenden Euler-Lagrange-Gleichungen (Lagrange-Gleichung 2. Art) mit einer ausführlichen Behandlung der Randterme und ihrer mechanischen Bedeutung. Die hinreichenden Bedingungen werden an den Legendreschen Bedingungen erörtert. Die aufgeführten Anwendungen enthalten die Brachistochrone, das Hamiltonsche Prinzip, Ausführungen zu Schwingungsproblemen, zum Rayleigh-Quotient und zu einigen Näherungsverfahren (harmonische Balance, Methode von Krylov-Bogoliubov). Ein kurzer Abriß über die Hamilton-Jacobi-Theorie der analytischen Dynamik und die Legendre-Transformation beschließt den Abschnitt. Das Kap. „Kontinuumsmechanik" gibt einen Überblick über Variationsprobleme der Elastizitätstheorie (Prinzip von Minimum der potentiellen Energie und der Komplementärenergie), der Strömungsmechanik und der Plastizitätstheorie. Es dient vor allem zur Systematisierung der verschiedenen Ansätze. Auf die Arbeit von J. B. Martin, H. Stumpf, A. D. I. Barr, K. Washizu, Yu-Chen u. a. wird eingegangen.
Die Literatur ist gebietsweise geordnet und ermöglicht eine gezielte Vertiefung des Stoffes.
Begrüßenswert sind auch die Notizen zur historischen Entwicklung der Variationsprinzipe.
Die vorliegenden CISM-Bände Nr. 54 und 65 können allen Lehrenden und Studierenden an Technischen Hochschulen empfohlen werden und stellen eine wertvolle Unterstützung für die Ausbildung dar.

Berlin B. Heimann

Rosemeier, G., Winddruckprobleme bei Bauwerken, Berlin-Heidelberg-New York, Springer-Verlag 1976, VII, 288 S., 171 Abb., DM 64,–. ISBN 3-540-07729-4

In dem Buch werden statische Windkraftprobleme, fremderregte stabile Bauwerkschwingungen (wie z. B. der Böeneffekt des natürlichen Windes und die Karmansche Nachlaufwirbelerregung) als auch durch aeroelastische Effekte bedingte selbsterregte Schwingungen (Galopping, Flatterschwingungen) behandelt. Der Autor versucht, die baustatisch wichtigen physikalischen Effekte und deren näherungsweise rechnerische Behandlung aufzuzeigen. Manches bleibt dabei allerdings oberflächlich, ungenau oder ist unrichtig. So muß es sehr verwundern, wenn z. B. das Lehrsche Dämpfungsmaß falsch eingeführt wird (S. 5), auf S. 75 Korrelationsfunktion und spektrale Dichte als ein und dieselbe Größe bezeichnet werden, die Frequenz f in der Spektraldichte $S(f)$ als „Frequenzband" bezeichnet wird, oder sogar bei der Gaußschen Normalverteilung nicht zwischen Argument und Funktionswert unterschieden wird (Formel (3.9)). Lineare Differentialgleichungen mit zeitabhängigen Koeffizenten nennt der Verfasser „stark nichtlinear" (S. 11).

Berlin A. Renger

Olsen, H. / Reinitzhuber, F. / Coblenz, A., Die zweiseitig gelagerte Platte. Die statische Berechnung von zweiseitig gelagerten rechteckigen Platten mit beliebigem Seitenverhältnis mittels Einfluß- und Zustandsflächen. Berlin-München-Düsseldorf, Verlag Wilhelm Ernst & Sohn 1979. VIII, 172 S., 80 Abb., 20 Tab., DM 80,–.

Das Buch stellt eine überarbeitete Neuauflage des zweibändigen Tabellenwerkes von Olsen/Reinitzhuber „Die zweiseitige gelagerte Platte" dar, das in den Jahren 1942–1962 in drei Auflagen erschienen ist. Da die zweiseitige Platte im konstruktiven Ingenieurbau sehr häufig vorkommt, war dieses Werk ein sehr willkommenes und viel benutztes Hilfsmittel für den Ingenieur. Aber auch heute, wo der Ingenieur der Praxis sich elektronischer Rechenanlagen bedient, dürfte ein Tabellenwerk dieser Art durchaus noch seine Berechtigung haben. Die theoretischen Grundlagen werden allgemein für orthotrope Platten dargelegt. Die Lösung der Plattengleichung erfolgt durch Reihenentwicklungen für Platten endlicher und unendlicher Breite mit gleichmäßiger Flächenbelastung, Einzellast an beliebiger Stelle sowie Linienlasten längs der freien Kanten. Die für eine Bemessung wichtigen Biegemomente werden am ausführlichsten behandelt. Für Punkte in Stützweitenmitte werden Einflußflächen angegeben, mit denen die maximalen Beanspruchungen für beliebige Belastungsfälle ermittelt werden können. Das Buch enthält dazu zahlreiche Diagramme und für eine große Zahl von Seitenverhältnissen Tabellen, deren Anwendung durch Zahlenbeispiele erläutert wird. Dabei werden neueste Belastungs- und Berechnungsvorschriften (DIN) berücksichtigt.

Wolterdorf G. Haberland

Kawata, K. / Shioiri, J. (Eds.), High Velocity Deformation of Solids. Symposium Tokyo/Japan August 24–27, 1977. International Union of Theoretical and Applied Mechanis (IUTAM). Berlin-Heidelberg-New York, Springer-Verlag 1978. XVIII, 452 S., 230 Abb., 20 Tab., DM 78.–, US $ 42,90. ISBN 3-540-09208-0

Der Band enthält alle 39 Vorträge, die auf Einladung auf dem IUTAM-Symposium über „Hochgeschwindigkeitsdeformation fester Körper" gehalten wurden, das unter Mitwirkung mehrerer wissenschaftlicher Gesellschaften Japans im Jahre 1977 in Tokio veranstaltet wurde. An diesem Symposium hatten 91 Wissenschaftler aus 9 Ländern teilgenommen, die überwiegende Mehrheit aus Japan. Etwa die Hälfte der Vorträge wurde von japanischen Wissenschaftlern gehalten, fast ein Drittel von Wissenschaftlern aus den USA. Die Vorträge behandelten vielfältige theoretische und experimentelle Untersuchungen makro- und mikromechanischer Eigenschaften von Metallen, Polymeren und Kompositwerkstoffen bei hohen Deformationsgeschwindigkeiten. Dabei werden solche Themen, wie die Dynamik der Dislokationen, die Wellenausbreitung, Deformation und das Bruchverhalten bei dynamischer Belastung sowie die Rißausbreitung vom Standpunkt der Mikromechanik behandelt. Weitere Themenkomplexe enthalten Vorträge zur Wellenausbreitung in anisotropen Medien, zum Materialverhalten bei kombinierter dynamischer Beanspruchung, zu elastisch-plastischen Wellen nach der Kontinuumstheorie der Dislokationen (Versetzungen), zur Wellenausbreitung in viskoelastischen Balken, zum Ultraschall- und schallnahen Stoß auf feste Körper. In anderen Themenkomplexen werden Vorträge zusammengefaßt über die Hochgeschwindigkeitsdeformation verschiedener Konstruktionselemente wie lange Stäbe, Schalen und Zylinder, über die Verbindung von Metallplatten durch Hochgeschwindigkeitskollision und über dynamische Reibungseffekte.

Der Band enthält auch die Disskusionsbeiträge.

Der Vorsitzende des Wissenschaftlichen Komitees, S. Fukui, hebt im Vorwort hervor, das Symposium habe deutlich gemacht, daß eine mikromechanische Behandlung oft notwendig ist für eine volles Verständnis des makroskopischen Verhaltens bei Hochgeschwindigkeitsdeformation und Bruch.

Zur Förderung der Verständnisses für die Zusammenhänge zwischen Mikrostruktur und mechanischem Verhalten wird somit auch dieser Tagungsband beitragen.

Woltersdorf G. Haberland

Decuyper, M., Exercices de mécanique eulérienne. Paris-New York-Barcelone-Milan, Masson 1980. 351 S., ISBN 2-225-65351-8

Der deutschsprachige Leser wird aus dem Titel der vorgelegten Aufgabensammlung nicht mit Sicherheit auf den Inhalt schließen können, da die gewählte Bezeichnung, „Übungen zur Eulerschen Mechanik", nicht der üblichen Terminologie entspricht.

Bei einer genauen Einsicht stellt man fest, daß in den beiden ersten Kapiteln eine allgemeine Einführung in die Vektor-Rechnung und die Kinematik eines starren Körpers gegeben wird. Alle Bezeichnungen werden allgemein räumlich abgehandelt (Richtungskosinus, Eulersche Winkel, Kardan-Winkel – hier Tait Bryan Winkel genannt – u. a. m.). In den folgenden 4 Kapiteln werden dann einfache, vorwiegend ebene, dynamische Aufgaben behandelt, wie sie dem Inhalt der Grundausbildung auf dem Gebiet der Technischen Mechanik für Ingenieure entsprechen. Das eigentliche Anliegen des Buches ist eine umfassende methodische und didaktische Aufbereitung dieses klassischen Lehrstoffes. Man kann der „Groupe des Rétroaction Didactique en Mécanique" bescheinigen, daß ihr dieses Anliegen gelungen ist.

Besonders hervorzuheben sind die sehr durchdachte Darstellung des Vektor-Kalküls im ersten Kapitel und die große Zahl sauberer, guter und anschaulicher Skizzen für die Übungsaufgaben. Sie sind für die gute Qualität der vorliegenden Aufgabensammlung mitbestimmend.

Der Lernende wird es auch sehr begrüßen, daß jedem Übungsabschnitt eine kurze „theoretische Einführung" vorangestellt wird. So erhält der Anfänger eine anschauliche und theoretisch klare Einführung in die Grundlagen der Dynamik starrer Körper, und der auf diesem Gebiet Lehrende kann manche methodischen und didaktischen Anregungen entnehmen.

Magdeburg J. Altenbach

Olszak, W. / Suklje, L. (Eds.), Limit Analysis and Rheological Approach in Soil Mechanics. CISM Courses and Lectures 217. Wien-New York, Springer-Verlag 1978. VII, 292 S., 116 Abb., DM 52,–. S 358,–. ISBN 3-211-81511-2

Einfache Lösungen der Plastizitätstheorie und der Rheologie sind auch in der Bodenmechanik seit langem bekannt. Aber erst in den letzten drei Jahrzehnten erfolgte eine durchgreifende Anpassung genereller Entwicklungen der Kontinuumsmechanik an die Besonderheiten des Materials „Lockergestein".

Das vorliegende Buch enthält 4 Vorlesungen eines Sommerkurses am Internationalen Zentrum für Mechanische Wissenschaften des Jahres 1976. J. De Jong stellt ein „Modell für das Verhalten granularer Materialien im Fließzustand" vor. Kopplungen der Gleichgewichtsbedingungen des ebenen Zustandes mit der Mohr-Coulombschen Hypothese als Bruchkriterium führt zu den besonders durch Caquot und Sokolovski bekannten Differentialgleichungen. Neu sind die Abschnitte 1.4. (Singularitäten im Feldesinneren) und 2. In letzterem Abschnitt beschreibt er ausgehend von den Grundgleichungen und ausschließlich rein kinemetrischen Betrachtungen ein von ihm entwickeltes sogenanntes „double sliding; free rotating"-Modell bis zur Formulierung der Randwertaufgabe. Z. Mroz und Cz. Szymanski beschäftigen sich mit „Nichtassoziierten Fließregeln zur Beschreibung des plastischen Fließens granularer Materialien"; als Fließbedingung wird ebenfalls die von Mohr-Coulomb benutzt. Eine hypothetisch gewählte Fließregel verbindet Spannungs- und Deformationszustand. Neben einer Relation, die den plastischen Anteil der Deformationsgeschwindigkeit allein dem Spannungszustand zuordnet, betrachten die Verfasser noch ein allgemeineres Fließgesetz, das einen Einfluß auch der Spannungsgeschwindigkeit voraussetzt. Sonderformen dieser Fließgesetze werden diskutiert, das Feld der Geschwindigkeiten und ein plastisches Fließen mit speziellen Annahmen für die Dilatanz besonders betrachtet. Hinweise für die Lösung der Randwertprobleme mit Beispielen (Stützwand, Ausfluß aus Silo schließen das Kapitel. J. Salençon gibt einen „Überblick über die Anwendung der Plastizitätstheorie in der Bodenmechanik". Er geht aus vom sukzessiven Fließen eines elasto-plastischen Materials, kommt dann zur Untersuchung des Grenzzustandes unter Beachtung der Inhomogenität des Bodens. Auch dieses Kapitel endet mit Beispielen und einer kritischen Analyse des Zutreffens und der Anwendbarkeit gewinnbarer Lösungen. L. Suklje schließlich schreibt über „Nichtlinear-viskoses Lockergesteinsverhalten". Er gibt einen ausgezeichneten Überblick über die modernen Entwicklungen und Vorstellungen in der Rheologie, die Interesse für die Bodenmechanik besitzen bzw. von Bodenmechanikern kreiert wurden. Besonders wertvoll ist die Darstellung der mehrdimensionalen Konsolidationstheorie sowie die Diskussion des Einflusses der verschiedenen, den Konsolidationsvorgang steuernden Parameter.

Das Buch ist sicher nicht als Lehrbuch mit einführendem Charakter gedacht; es ist nicht für den Praktiker geschrieben, der Hinweise auf die Anwendung zur Lösung seiner speziellen Aufgaben erwartet. Auch der mit der Problematik besser Vertraute hat gewisse Schwierigkeiten beim Lesen, die schon aus unterschiedlichen Bezeichnungen in den verschiedenen Kapiteln resultieren. Der Beitrag von Salençon ist zudem in französischer Sprache verfaßt, die der anderen Autoren in englisch.

Der theoretisch Arbeitende findet in dem Buch eine gute Darstellung der Grenzzustandstheorien unter verschiedenen Aspekten und letzter Entwicklungen in der Rheologie der Lockergesteine. Ihm ist es zu empfehlen.

Freiberg W. Förster

Leipholz, H. H. E. (Ed.), Structural Control. Proceedings of IUTAM Symposium, Waterloo, Canada, 1979. Amsterdam-New York, North-Holland Publ. Co. 1980. 822 S., US $ 87.75. Dfl. 180.00. ISBN 0444-85485-1

The volume contains 43 papers concerning structural optimization and structural control. The main objective of the meeting concerned the structural control which roughly speaking combines optimal design and control theory to arrive at desired structural behaviour under expected loadings. The problem consists in assuring a desired deformability, required resistance, controlled oscillations of structures belonging to aeronautical as well as to civil engineering under wind load and earthquakes, high speed aircrafts of controlled flutter, high rise building with constrained deformability and wind induced oscillations constitute examples of such structures. The control can be passive when suitable devices are incorporated in a structure like prestressing additional masses, or active when appropriate sensors often via on line operating computers activate deviced actuators giving rise to forces reducing the undesirable effects of excessive deformation, accelerations, flutter or vibrations.

Three groups of papers can be distinguished in the volume. The first one is related to the control theory and its application to structural optimization. The second one concerns also mathematical aspects but this time of the structural control against undesirable effects. The last deals with specific engineering examples of structural control both passive and active from aeronautics, high rise buildings and bridges.

Some papers constitute the state-of-the-art reports written by specialists either in the optimum control theory or creative engineers. A number of papers concerns specific new solutions. The reviewer feels worthwhile to mention the contributors like R. E. KLEIN, J. M. KELLY, H. H. E. LEIPHOLZ, U. LEPIK, Z. MRÓZ, J. ROORDA, T. T. SOONG, J. T. P. YAO, W. ZUK.

The book reflects the aims of the symposium to get together for an exchange of concepts, results and experience two groups of people: mainly researchers concerned with fundamentals of the control theory as well as researchers and engineers involved in applications of the active and passive control both to civil and to aeronautical structures.

The volume contains a very valuable information about the present State of knowledge and engineering in this new domain of structural mechanics and engineering termed the structural control. No doubt that this domain will grow in importance due to the tendency to develop lighter less rigid structures or energy absorbing devices. It requries both the knowledge of structural mechanics and of control theory involving as well the sense of engineering. The book should thus interest theroreticians and structural engineers and contains an inspiring material supplemented by references to the sources.

Warsaw A. SAWCZUK

Болотин, В. В./ Новичков, Ю. Н., Механика многослойных конструкций. Москва, Машиностроение 1980. 375сс., 3р. 20К.

Bolotin, W. W. / Nowitschkow, Ju. N., Mechanik vielschichtiger Konstruktionen. (In Russisch). Moskau, Maschinenbau 1980, 375 S., 3,20 Rubel.

In dem vorliegenden Buch werden Berechnungsmethoden für vielschichtige Konstruktionen bei statischer und dynamischer Belastung, bei Temperatureinwirkung und bei Stabilität vorgestellt, Methoden zur Bestimmung effektiver physikalisch-mechanischer Kenngrößen beschrieben und Hinweise auf die optimale Auswahl von Parametern geschichteter Verbundwerkstoffe gegeben. Das Buch betrachtet eine Richtung der Mechanik, deren Entwicklung von den Autoren aktiv betrieben und beeinflußt wurde.

Im Kapitel 1 werden die Grundlagen der Theorie vielschichtiger Konstruktionen dargelegt. Sie erstrecken sich auf die Herleitung der Grundgleichungen des Gleichgewichts, der Schwingungen und der Stabilität sowie der natürlichen Randbedingungen aus den Variationsprinzipien von LAGRANGE, HAMILTON-OSTROGRADSKIJ und TREFFTZ und auf die Klassifizierung der verschiedenen Schichten des Verbundkörpers bezüglich ihrer mechanischen Kenngrößen in tragende (steife) und verbindende (weiche) Schichten. Verallgemeinerungen hinsichtlich Anisotropie und Berücksichtigung des Querkraftschubes in den tragenden Schichten runden diesen Abschnitt ab. Das Kapitel 2 bringt die Methoden der Theorie vielschichtiger Konstruktionen. Die Grundgleichungen dieser Verbundkörper sind Differentialgleichungen bezüglich der Koordinaten in den Schichtenebenen und haben Differenzcharakter hinsichtlich der dritten Koordinate längs der Normalen zu den Schichten, die abhängig von der Schichtennumerierung diskrete Werte annimmt. Einen gewissen Schwerpunkt bilden die mathematischen Lösungsverfahren für Konstruktionen, z. B. mit regelmäßigen Schichtenaufbau über die Wanddicke, für die geschichtete Platte und den parallel und orthogonal zu seiner Oberfläche geschichteten Halbraum, deren Lösungen die Anzahl der Schichten nur noch als Parameter ausweisen. Im Kap. 3 wird die Theorie der vielschichtigen Konstruktionen mit krummlinigen Schichten vorgestellt. Sie bringt die Erweiterung der im Kap. 1 angeführten Gedankengänge für ebene Schichten auf krummlinige. Diese Grundlagen finden im Kap. 4 bei der Berechnung vielschichtiger Schalen unter statischer Belastung, vorgeführt bei Zylinder- und Kugelschalen mit regelmäßiger Schichtenstruktur, Anwendung. Das Kap. 5 ist Temperaturaufgaben bei vielschichtigen Konstruktionen gewidmet. Zunächst erfolgt die Bestimmung der Temperaturfelder über die Formulierung des Variationsprinzips der Wärmeleitung, anschließend kommen die Formulierung der Gleichungen der Thermoelastizität, das Vorstellen von Lösungsmethoden und ihre Anwendung auf vielschichtige Zylinderschalen. Das Kap. 6 ist der geometrisch nichtlinearen Theorie und Stabilitätsaufgaben vorbehalten. Es werden die im Kap. 3 aufgestellten Schalengleichungen erweitert, Möglichkeiten ihrer Linearisierung diskutiert und auf Stabilitätsprobleme bei speziellen vielschichtigen Konstruktionen angewendet. Im Mittelpunkt des Kap. 7 steht die dynamische Berechnung vielschichtiger Konstruktionen. Schwerpunkte sind die Diskussion von Dämpfungsmodellen, die Betrachtung von Eigenschwingungen und erzwungenen Schwingungen und die aktive Einflußnahme auf das Schwingungsverhalten mittels schwingungsdämpfender Oberflächenhüllen. Die Kapitel 8 und 9 machen den Leser mit den Grundlagen und technologischen Aufgaben der Mechanik der Verbundwerkstoffe bekannt. Interessant ist hier vor allem das Prinzip der energetischen Kontinualisation, der Ersatz des heterogenen Verbundwerkstoffes durch äquivalentes homogenes Material, entwickelt aus der Theorie der vielschichtigen Platte.

Das Buch wendet sich an wissenschaftliche Mitarbeiter, Berechnungsingenieure und Konstrukteure sowohl des Maschinenbaus als auch des Bauwesens. Die Form der Darstellung, die Bereitstellung wesentlicher mathematischer Grundlagen zu den entsprechenden Kapiteln und die Quantifizierung vieler theoretischer Beziehungen in Form von Anwendungsbeispielen erleichtern das Studium des Buches und lassen seine Schwerpunkte klar erkennen.

Dresden D. WITT

de Groot, S. R. / Mazur, P., Anwendung der Thermodynamik irreversibler Prozesse, Mannheim-Wien-Zürich, B. I.-Wissenschaftsverlag 1974, 349 S., ISBN 3-411-01449-0

Bekanntlich kann in einem adiabatisch isolierten System die Abweichung ΔS der Entropie von ihrem Wert im Gleichgewichtszustand durch einen quadratischen Ausdruck in den Zustandsvariablen beschrieben werden. Wenn man annimmt, das zeitliche Verhalten der Zustandsparameter werde phänomenologisch durch ein homogenes lineares Differentialgleichungssystem mit konstanten Koeffizienten beschrieben, so gelten für seine Koeffizienten die ONSAGERschen Reziprozitätsbeziehungen.

Ausgehend von ihren „Grundlagen der Thermodynamik irreversibler Prozesse", wo sie die Reziprozitätsgesetze aus der zeitlichen Umkehrbarkeit der Mikroprozesse abgeleitet hatten, behandeln die Autoren im vorliegenden Buch die Entropiebilanz für die häufigsten in der Praxis auftretenden irreversiblen Prozesse: Chemische Reaktionen, Wärmeleitung, Diffusion, zähe Strömungen, elektrischen Strom, irreversible Prozesse in polarisierten Medien und diskontinuierliche Systeme.

Berlin J. FÖRSTE

Meetz, K. / Engl, W. L., Elektromagnetische Felder. Mathematische und physikalische Grundlagen. Anwendungen in Physik und Technik. Hochschultext. Berlin-Heidelberg-New York, Springer-Verlag 1980. 192 Abb., XIII, 668 S., DM 69,–. US $ 40.80. ISBN 3-540-09597-7

Mit diesem Buch liegt eine Darstellung der Theorie des elektromagnetischen Feldes vor, in der die Feldgrößen als Differentialformen erscheinen. Die Gesetze des elektromagnetischen Feldes werden durch Relationen zwischen diesen Differentialformen beschrieben, wobei geometrische Eigenschaften des Raumes durch Materialgleichungen bedingt sind, welche Intensitäts- und Qualitätsgrößen verbinden. In den beiden ersten Kapiteln werden die erforderlichen mathematischen Grundlagen entwickelt. Danach werden behandelt das elektrische Feld ruhender Ladungen und Randwertaufgaben der Potentialtheorie, ferner das magnetische Feld stationärer Ströme mit zugehörigen Randwertaufgaben. Es folgt die maxwellsche Theorie des elektromagnetischen Feldes. Die Materialeigenschaften werden im 8. Kap. behandelt: Materialgleichungen für metallische Leiter, Dispersionsrelationen der Materialkonstanten, Feldgleichungen und Materialgleichungen für bewegte Medien, elektrische und magnetische Feldkräfte. Das 9. Kap. bringt Ausbreitungserscheinungen der elektromagnetischen Wellen, KLEIN-GORDON-Gleichung, Telegrafengleichung, Reflektion und Absorbtion, Oberflächenwellen, Ausbreitung in einem dispersiven Halbraum, Zylindersymmetrische Probleme, Wellenleiter, Ausstrahlung elektromagnetischer Wellen, Reziprozitätstheorem. Im 10. Kap. wird die Theorie der Gleich- und Wechselstromnetzwerke entwickelt; die hierzu erforderlichen Grundlagen der Graphentheorie findet man in einem Anhang am Ende des Buches. Das 11. Kap. behandelt die spezielle Relativitätstheorie.

In einem letzten Kapitel wird die Wechselwirkung zwischen bewegten Ladungen und Feldern beschrieben: Das Feld einer bewegten Punktladung, LIENARD-WIECHERT-Potentiale für gleichförmig bewegte und für beschleunigte Punktladungen, Selbstwechselwirkung, LORENTZ-DIRAC-Gleichung, Wechselwirkung mehrerer Ladungen. Als Wechselwirkungsprozesse werden betrachtet: Die Ablenkung von Teilchenstrahlen in Feldern, die Streuung von Wellen an Teilchen und das Entstehen elektromagnetischer Strahlung beim Abbremsen von Teilchen in Feldern. Voraussetzungen für das Verständnis sind bei diesem Buch allgemeine Kenntnisse der analytischen Geometrie und der Differential- und Integralrechnung. Das Buch ist zum Studium der theoretischen Physik als Lehrbuch über Elektrodynamik zu empfehlen.

Berlin E. HEYN

Krause, F. / Rädler, K.-H., Mean-Field Magnetodynamics and Dynamo Theory. Berlin, Akademie-Verlag 1980. 271 S., 46 Abb., M 48,–. BN 761 626 7 (5948)

Man ist heutzutage davon überzeugt, daß die beobachteten Magnetfelder von Erde, Sonne und anderen Himmelskörpern in erster Linie auf Strömungen elektrisch leitfähiger Flüssigkeiten oder Gase im Innern dieser Körper zurückzuführen sind. Deswegen bilden die magnetohydrodynamischen Grundgleichungen den Ausgangspunkt aller Versuche einer theoretischen Begründung. Wegen der mathematischen Schwierigkeiten, welche die nichtlineare Wechselwirkung zwischen den hydrodynamischen und elektromagnetischen Feldern bereitet, wird die Rückwirkung des Magnetfeldes auf das Geschwindigkeitsfeld in geo- und astrophysikalischen Untersuchungen zur Dynamotheorie gewöhnlich vernachlässigt. Man macht sich anschauliche Vorstellungen über die Geschwindigkeitsverteilung und bestimmt daraus das Magnetfeld.

Die Theorie der in der vorliegenden Monographie behandelten mittleren Felder geht von der Überlegung aus, daß sich alle in die magnetohydrodynamischen Grundgleichungen eingehenden Felder in Mittelwert und Schwankungsgröße zerlegen lassen. Nachdem man in den Gleichungen zu Mittelwerten übergegangen ist, sucht man die gemittelten elektromagnetischen Feldgrößen aus bekannt angenommenen Mittelwerten von Stromdichte und Geschwindigkeit sowie gewissen Eigenschaften der Schwankungsgeschwindigkeit zu bestimmen. Als entscheidende Einflußgröße wird dabei die sogenannte turbulente elektromotorische Kraft angesehen; für diese wird ein linearer Ansatz in der mittleren magnetischen Induktion $\overline{\mathfrak{B}}$ und ihren ersten Ortsableitungen gemacht. Dieser Ansatz wird auf verschiedene Weise spezialisiert; auf mathematischem Wege werden Schlußfolgerungen gezogen. Auf dieser Grundlage gelingt es, das Magnetfeld der Erde und die Schmetterlingsstruktur der Sonnenfleckenzyklen größenordnungsmäßig richtig wiederzugeben.

Die Darstellung basiert auf langjähriger Forschung im Zentralinstitut für Astrophysik in Potsdam und früher im Institut für Magnetohydrodynamik in Jena.

Berlin J. FÖRSTE

Bullough, R. K. / Caudrey, P. J. (Eds.), Solitons. Topics in Current Physics 17. Berlin-Heidelberg-New York, Springer-Verlag 1980. XVIII, 389 S., 20 Abb., DM 76,–. US $ 44.90. ISBN 3-540-09962-X

Zu der Thematik findet man in der Literatur einige Tagungsbände wie die früher an dieser Stelle besprochenen (s. ZAMM *60*, Nr. 3 und Nr. 10 (1980)), und man findet das empfehlenswerte Lehrbuch von ZAKHAROV u. a. (Nauka, Moskau (1980)). Das vorliegende Buch ist weder Tagungsband noch Lehrbuch. Es wurde konzipiert im Jahre 1975 (Redaktionsschluß 1979) und bietet eine repräsentative Darstellung des Standes der Solitonentheorie zu dieser Zeit. Die bekanntesten damals auf diesem Gebiet arbeitenden Gruppen sind vertreten. So finden wir als Autoren BULLOUGH, CAUDREY und GIBBS (England); ZAKHAROV, FADDEEV und NOVIKOV (UdSSR); LAMB, MCLAUGHLIN und NEWELL (USA); HIROTA, TODA und WADATI (Japan); CALOGERO und DEGASPERIS (Italien) und LUTHER (Dänemark).

Der Leser findet einen Zugang zu der und eine Einsicht in die Theorie der Solitonen aus der Sicht verschiedener Schulen. Die Beiträge sind im wesentlichen zusammenfassende Darstellungen bereits veröffentlichter Arbeiten. Dagegen zeichnen sich die Artikel von Newell und von Zakharov durch neue methodische Gesichtspunkte und daraus resultierende neue Ergebnisse aus.

Abgesehen von zwei anwendungsorientierten Artikeln sind systematisch-methodische Gesichtspunkte vorherrschend. Von der Quantentheorie der Solitonen werden nur einige Aspekte behandelt in den Beiträgen von FADDEEV und von LUTHER. In einem einleitenden, gut lesbaren Artikel behandeln die Herausgeber die Geschichte des Solitons vom Jahre 1834 an und gehen dabei in einem abschließenden Paragraphen auch auf wichtige Fortschritte und neue Gesichtpunkte ein, die in den Jahren 1975–1979 hervorgetreten sind und sonst in dem Band nicht behandelt werden. Trotz der anhaltenden stürmischen Entwicklung der Solitonentheorie darf angenommen werden, daß der vorliegende Sammelband auf längere Zeit seinen Platz als Standardwerk behaupten wird.

Berlin H. STEUDEL

Landau, L. D. / Lifschitz, E. M., Lehrbuch der Theoretischen Physik. Bd. 9: Statistische Physik, Teil 2 (von **Lifschitz, E. M. / Pitajewski, L. P.**). Berlin, Akademie-Verlag 1980. XI, 390 S., M 28,. BN 7 623 184 (5436/IX). Übersetzung aus dem Russischen

Der vorliegende zweite Teil der Neuauflage „Statistische Physik" befaßt sich mit der Quantentheorie der kondensierten Materie. Ausgangspunkt bildet die Theorie der Quantenflüssigkeiten (der in der ersten Auflage lediglich 3 Paragraphen gewidmet waren), da die hier entwickelten Begriffsbildungen die Grundlage für die quantentheoretische Beschreibung makroskopischer Körper bilden. Eine besondere Rolle spielt dabei die Methode der Greenschen Funktionen, der in den Kapiteln II, III, IV, V, VI und VIII ausführliche und den jeweiligen Problemen entsprechende Darstellungen gewidmet sind; dazu gehören auch die eingehenden Erläuterungen der zugeordneten Diagrammtechniken. Insbesondere bietet die Einführung in diese Methodik in Kapitel II eine nicht nur für die speziellen Probleme dieses Buches nützliche Ergänzung zu den entsprechenden Kapiteln VIII und XI in Band IV. Das besondere Gewicht, das der Methode der Greenschen Funktionen zugemessen wird, hat u. a. auch dazu geführt, daß die ursprünglich in den Bänden VIII und VI befindlichen Kapitel über elektromagnetische und hydrodynamische Fluktuationen jetzt in den vorliegenden Band, und zwar erheblich erweitert, übernommen wurden. Wie alle übrigen Bände enthält auch dieser Band wieder eine grosse Zahl von Übungsaufgaben mit Lösungen – einer von vielen didaktischen Vorzügen dieses Lehrbuches. Die Kapitel des Buches sind: Die normale Fermi-Flüssigkeit; Greensche Funktionen eines Fermi-Systems bei $T = 0$, Die Superfluidität, Greensche Funktionen bei endlichen Temperaturen, Die Supraleitfähigkeit, Elektronen im Kristallgitter, Der Magnetismus, Elektromagnetische Fluktuationen, Hydrodynamische Fluktuationen.

Gießen G. BRAUNSS

Haken, H., (Ed.), Dynamics of Synergetic Systems. Proceedings of the International Symposium on Synergetics. Bielefeld, September, 1979. Springer Series in Synergetics 6, Berlin-Heidelberg-New York, Springer-Verlag 1980. VIII, 271 S., 146 Abb., DM 68,–. US $ 40.20. ISBN 3-540-09918-2

Die Beiträge in dem vorliegenden sechsten Band der Synergetics-Reihe des Springer-Verlages sind der Untersuchung vorwiegend kooperativer Effekte komplexer Systeme in verschiedenen naturwissenschaftlichen Bereichen gewidmet, die zu makroskopischen räumlichen, zeitlichen oder funktionellen Strukturen führen. Dabei werden sowohl deterministische als auch stochastische Prozesse betrachtet. Im Rahmen dieser Untersuchungen spielt die Bifurkationstheorie dynamischer Systeme eine zentrale Rolle, wie bereits in der ausführlichen Einleitung von H. HAKEN am Beispiel des Lasers demonstriert wird. In diesem Beitrag wird außerdem über ein neues synergetisches Phänomen in der Lasertheorie berichtet. Von den Arbeiten zu der Gleichgewichtsphasenübergängen sei der Übersichtsartikel von H. E. STANLEY et al. hervorgehoben. Hinsichtlich der Untersuchungen auf dem Gebiet der Nichtgleichgewichtsphasenübergänge verweisen wir auf den Beitrag von W. HORSTHEMKE über den Einfluß äußerer stochastischer Fluktuationen. Der Band enthält weiterhin Artikel von M. L. SMOES, P. C. FIFE, L. ARNOLD und A. NITZIAN über die räumlich-zeitliche Organisation chemischer Prozesse auf makroskopischer Ebene. Zwei Beiträge von P. SCHUSTER und K. SIGMUND sind dem Problemkreis der Selbstorganisation biologischer Makromoleküle gewidmet. Von den Untersuchungen zur Dynamik von Systemen mit zahlreichen Teilsystemen seien die Arbeiten von H. G. OTHMER und R. LEFÉVER genannt. Im Beitrag von J. S. NICOLIS über kognitive Netzwerke ist eine interessante Kombination von Spieltheorie und Kombinationstheorie enthalten. Neben den Untersuchungen zur Mastergleichung (J. WM. TURNER, C. W. GARDINGER) sei noch auf die Beiträge von Y. KURAMOTO und O. E. RÖSSLER zum Chaos- und Turbulenzproblem verwiesen. Für fast alle Artikel ist charakteristisch, daß sie offene Probleme aufzeigen und zu weiteren Forschungen anregen.

Berlin K. R. SCHNEIDER

Izé, A. F. (Ed.), Functional Differential Equations and Bifurcation. Proceedings, São Carlos, Brazil 1979. Lecture Notes in Mathematics 799. Berlin-Heidelberg-New York, Springer-Verlag 1980. XXII, 409 S., 21 Abb., DM 43,50. US $ 25.70. ISBN 3-540-09986-7

Der vorliegende LNM-Band enthält 21 Vorträge, die auf der Konferenz über Funktional-Differential-Gleichungen und Bifurkationstheorie gehalten wurden, die Anfang Juli 1979 in São Paulo stattfand. Zentrales Anliegen aller Vorträge ist das Studium des qualitativen Verhaltens der Lösungen von Funktional-Differentialgleichungen (vorwiegend gewöhnliche Differentialgleichungen mit nacheilendem Argument), wobei in einigen Fällen der Einfluß von Parametern berücksichtigt wird (Bifurkationsprobleme). Die folgenden Vorträge charakterisieren das Spektrum der Vorträge recht gut: CASTELAN, W., A Liapunov functional for a matrix retarded difference-differential equation with several delays; SHOW, S.-N., Periodic solutions of nonlinear autonomous hyperbolic equations; HALE, J. K., Some recent results on dissipative processes; IZÉ, A. F. and VENTURA, A., Relationship in the neighbourhood of infinity and asymptotic equivalence of neutral functional differential equations; MAGNUS, R., Topological equivalence in bifurcation theory; MILOJEVIĆ, P. S., Approximation — solvability of some nonlinear operator equations with applications; OLIVEIRA, M. C., Non-singular structural stable flows on three-dimensional manifolds; RUIZ-CLAEYSSEN, J. and COCKBURN, B., Applications of the integral averaging bifurcation method to retarded functional differential equations; TAKENS, F., Moduli and bifurcations; non-transversal intersections of invariant manifolds of vectorfields; YOSHIZAWA, T., Stability properties in almost periodic systems of functional differential equations.

Berlin K. R. SCHNEIDER

Kall, P. / Prekopa, A. (Eds.), Recent Results in Stochastic Programming, Proceedings, Oberwolfach 1979, Lecture Notes in Economics and Mathematical Systems 179, Berlin-Heidelberg-New York, Springer-Verlag 1980, IX, 237 S., 10 Abb., DM 34,–, US $ 20.10. ISBN 3-540-10013-X

This volume is divided into two parts of quite different character. While the first one contains papers presenting theoretical results the papers in part II treat mainly of stochastic programming. — Two papers of part I are concerned with continuity and differentiability in parametric linear programming (BOL, BROSOWSKI), another shows the close relations of parametric and stochastic programming (BEREANU). All are more or less direct continuations of former papers of the respective authors. The three papers by GROENEWEGEN and WESSELS, HEILMANN, KLEIN HANEVELD study programming problems via dynamic programming. KLEIN HANEVELD's paper especially considers inventory control problems, a classical domain for applying dynamic and stochastic programming. The last paper in part I (RINOTT) demonstrates applications of mathematical programming to statistics. — The most papers of part II concern applications of results on stochastic programming obtained by PRÉKOPA and his coworkers to economic and technical problems like water reservoirs, energy resources etc. (DÉAK, DUPAČOVÁ, KARREMAN, KELLE, PRÉKOPA). K. MARTI's paper presents an approach to solve stochastic linear programs by semi-stochastic approximation and compares it with numerical results obtained by other authors. KALL's extended abstract also refers to computational methods in stochastic linear programming, while the paper by KALLBERG and ZIEMBA on application of an extended FRANK-WOLFE algorithm to portfolio selection has at most indirect connections to the subject mentionned in the title of the book. — Studying the papers of the present volume in total the reader is impressed by the influence PRÉKOPA's concept of logarithmic concave measure has had and still has to the development of stochastic programming and its applications.

Berlin H. WEINERT

In der DDR beziehbar im Rahmen des Kontingents.

EINGEGANGENE BÜCHER

Bei der Schriftleitung sind folgende Bücher eingegangen (ausführliche Besprechung bleibt vorbehalten).

Prabhu, N. U., Stochastic Storyge Processes, Queues, Insurance Risk, and Dams, Applications of Mathematics 15, Berlin-Heidelberg-New York, Springer-Verlag 1980, XII, 140 S., DM 38,–. US $ 22.50. ISBN 3-540-90522-7

Mercier, B., Lectures on Topics in Finite Element Solution of Elliptic Problems, Notes by Vijayasundaram, G., Berlin-Heidelberg-New York, Springer-Verlag and Bombay, Tata Institute of Fundamental Research 1979, V, 191 S., 30 Abb., DM 16,–. US $ 9.50. ISBN 3-540-09543-8

Segel, L. A. (ed.), Mathematical Models in Molecular and Cellular Biology, X, 757 S., Cambridge, Cambridge University Press 1980, ISBN 0-521-22925-1

Jones, N. D. (ed.), Semantics-Directed Compiler Generation, Proceedings of a Workshop, Aarhus, Denmark, 1980, Lecture Notes in Computer Science 94, Berlin-Heidelberg-New York, Springer-Verlag 1980. V, 489 S., 34 Abb., DM 45.50. US $ 26.90. ISBN 3-540-10250-7

Has'minskii, R. Z., Stochastic Stability of Differential Equations. Mechanics: Analysis 7. Alphen aan den Rijn, Sijthoff & Noordhoff 1980. 360 S., Dfl. 100.00. $ 50.00. ISBN 90-286-0100-7

Beauville, A. (ed.), Algebraic geometry, Angers, 1979. Alphen aan den Rijn, Sijthoff & Noordhoff 1980. 324 S., Dfl. 100.00. $ 50.00 ISBN 90-286-0500-2

Milner, R., A Calculus of Communicating Systems, Lecture Notes in Computer Science 92, Berlin-Heidelberg-New York, Springer-Verlag 1980, VI, 171 S., DM 21,50. US $ 12.70. ISBN 3-540-10235-3

Nijholt, A., Context-Free Grammars: Convers, Normal Forms, and Parsing, Lecture Notes in Computer Science 93; Berlin-Heidelberg-New York, Springer-Verlag 1980, VII, 253 S., 29 Abb., DM 29,50. US $ 17.40. ISBN 3-540-10245-0

Marlin, C. D., Coroutines — A Programming Methodology, a Language Design and an Implementation, Lecture Notes in Computer Science 95, Berlin-Heidelberg-New York, Springer-Verlag 1980, XII, 246 S., 55 Abb., DM 29,50. US $ 17.40. ISBN 3-540-10256-6

Bjorner, D. /Oest, O. N. (eds.), Towards a Formal Description of Ada. Lecture Notes in Computer Science 98. Berlin-Heidelberg-New York, Springer-Verlag 1980. XIV, 630 S., DM 46,—. US $ 27.20. ISBN 3-540-10283-3

Bosznay, A. (ed,), Bracketing of Eigenfequencies of Continuous Structures, Euromech Colloquium 122, Mátrafüred (Hungary), 1979, Budapest, Akadémiai Kiadó 1980, 669 S., Ft 480,—. ISBN 963-05-2625-5

Lamb, G. L., jr., Elements of Soliton Theory, New York-Chichester-Brisbane-Toronto, John Wiley & Sons 1980, XII, 289 S., £ 16.00. ISBN 0-471-04559-4

Encarnacao, J. (ed.), Computer Aided Design Modelling Systems Engineering, CAD-Systems, CREST Advanced Course, Darmstadt 1980, Lecture Notes in Computer Science 89, Berlin-Heidelberg-New York, Springer-Verlag 1980. XIV, 461 S., 219 Abb., DM 45,50. US $ 26.90. ISBN 3-540-10242-6

Martini, R. (ed.), Geometrical Approaches to Differential Equations, Proceedings, Scheveningen, The Netherlands 1979, Lecture Notes in Mathematics 810, Berlin-Heidelberg-New York, Springer Verlag 1980, VII, 339 S., 64 Abb., DM 34,50. US $ 20.40. ISBN 3-540-10018-0

Gumowski, I. / Mira, C., Recurrences and Discrete Dynamic Systems, Lecture Notes in Mathematics 809, Berlin-Heidelberg-New York, Springer-Verlag 1980, VI, 272 S., 114 Abb., DM 29,—. US $ 17.20. ISBN 3-540-10017-2

Rizzi, A. / Viviand, H. (eds.), Numerical Methods for the Computation of Inviscid Transonic Flows with Shock Waves, Notes on Numerical Fluid Mechanics 3, Braunschweig, Vieweg 1981, XVI, 266 S., 121 Abb., DM 72,—. $ 44.00. ISBN 3-528-08077-9

Fenyö, S., Moderne mathematische Methoden in der Technik, Bd. 3, ISNM 18. Basel-Boston-Stuttgart, Birkhäuser Verlag 1980, 347 S., sfr. 84.—, DM 92,—. ISBN 3-7643-1097-9

Jacobson, D. H. / Martin, D. H. / Pachter, M. / Geveci, T., Extensions of Linear-Quadratic Control Theory. Lecture Notes in Control and Information Sciences 27, Berlin-Heidelberg-New York, Springer-Verlag 1980. XI, 288 S., 6 Abb., DM 32.50, US $ 19.20. ISBN 3-540-10069-5

Boiti, M. / Pempinelli, F. / Soliani, G. (eds.), Nonlinear Evolution Equations and Dynamical Systems, Proceedings, Lecce 1979 Lecture Notes in Physics 120, Berlin-Heidelberg-New York, Springer-Verlag 1980, VI, 368 S., 5 Abb., DM 43,—, US $ 25.40. ISBN 3-540-09971-9

Gaier, D., Vorlesungen über Approximation im Komplexen, Basel-Boston-Stuttgart, Birkhäuser Verlag 1980, 173 S., sFr. 26.—, DM 28,—. ISBN 3-7643-1161-4

Fraeijs de Veubeke, B. M., Memorial Volume of Selected Papers, Solid Mechanics Archives, Alphen aan den Rijn, Sijthoff & Noordhoff 1980, and Waterloo, Solid Mechanics Division, Waterloo University, 1980, XVII, 752 S., ISBN 90-286-0900-8

Meirovitch, L., Computational Methods in Structural Dynamics, Alphen aan den Rijn, Sijthoff & Noordhoff 1980, XIV, 439 S., ISBN 90-286-0580-0

Osaki, S. / Nishio, T., Reliability Evalution of Some Fault-Tolerant Computer Architectures. Lecture Notes in Computer Science 97. Berlin-Heidelberg-New York, Springer-Verlag 1980. VI, 129 S., 46 Abb., DM 18,—. US $ 10.70. ISBN 3-540-10274-4

Peterson, J. L., Computer Programs for Spelling Correction: An Experiment in Program Design. Lecture Notes in Computer Science 96. Berlin-Heidelberg-New York, Springer-Verlag 1980. VI, 213 S., 8 Abb., DM 25,—. US $ 14.80. ISBN 3-450-10259-0

Buchberger, B. / Lichtenberger, F., Mathematik für Informatiker I, Die Methode der Mathematik. Informatik-Fachberichte 35. Berlin-Heidelberg-New York, Springer-Verlag 1980. XI, 315 S., DM 33,—. US $ 19.50. ISBN 3-540-10417-8

Riddle, W. E. / Fairley, R. E., Software Development Tools. Berlin-Heidelberg-New York, Springer-Verlag 1980. VIII, 280 S., 26 Abb., DM 39.50 US $ 23.30. ISBN 3-540-10326-0

Libby, P. A. / Williams, F. A. (eds.), Turbulent Reacting Flows, Topics in Applied Physics 44, Berlin-Heidelberg-New York, Springer-Verlag 1980, XI, 243 S., 38 Abb., DM 84,—. US $ 49.60. ISBN 3-540-10192-6

Hinton, E. / Owen, D. R. J., Finite Element Programming, Computational Mathematics and Applications, London, Academic Press, 1977, XII, 305 S., $ 18.00 ISBN 0-12-349352-8 p/b

Engels, H., Numerical Quadrature and Cubature, Computational Mathematics and Applications, London, Academic Press 1980, XIV, 414 S., $ 74.00. ISBN 0-12-238850-X

Gierz, G. u. a., A Compendium of Continuous Lattices, Berlin-Heidelberg-New York, Springer-Verlag 1980. XX, 371 S., DM 38,—. US $ 22.50. ISBN 3-540-10111-X

Krylov, N. V., Controlled Diffusion Processes, Applications of Mathematics 14, Berlin-Heidelberg-New York, Springer-Verlag 1980, XII, 308 S., DM 79,—. US $ 44.30. ISBN 3-540-90461-1. (Translation from the Russian)

Bradbury, L. J. S. u. a. (eds.), Turbulent Shear Flows 2, Selected Papers from the Second International Symposium on Turbulent Shear Flows, Berlin-Heidelberg-New York, Springer-Verlag 1980, IX, 391 S., 310 Abb., DM 128,—. US $ 75.60. ISBN 3-540-10067-9

Truesdell, C. III, The Tragicomical History of Thermodynamics, 1822—1854, Studies in the History of Mathematics and Physical Sciences 4, Berlin-Heidelberg-New York, Springer-Verlag 1980. XIII, 372 S., 8 Abb., DM 99,—, US $ 58.50. ISBN 3-540-90403-4

Kallianpur, G., Stochastic Filtering Theory, Applications of Mathematics 13, Berlin-Heidelberg-New York, Springer-Verlag 1980, XVI, 316 S., DM 59.50, US $ 35.10. ISBN 3-540-90445-X

McClamroch, N. H., State Models of Dynamic Systems, A Case Study Approach, Berlin-Heidelber-New York, Springer-Verlag 1980, VIII, 248 S., 97 Abb., DM 34,—, US $ 20.00 ISBN 3-540-90490-5

Collatz, L. / Meinardus, G. / Wetterling, W., (eds.), Konstruktive Methoden der finiten nichtlinearen Optimierung, Tagung, Oberwolfach, 1980, ISNM 55, Basel-Boston-Stuttgart, Birkhäuser Verlag 1980, 211 S., sFr. 44.— —. ISBN 3-7643-1205-X

Bandemer, H. / Näther, W., Theorie und Anwendung der optimalen Versuchsplanung II, Handbuch zur Anwendung, Mathematische Monographien 48, Berlin, Akademie-Verlag 1980, XII, 590 S., 86,— M, BN 7623192 (6215/2)

Kovács, L. B., Combinatorial Methods of Discrete Programming, Mathematical Methods of Operations Research 2, Budapest, Akadémiai Kiado 1980, 283 S., ISBN 96305200-4

Calladine, C. R., Engineering Plasticity, Oxford-London, Pergamon Press 1969, 328 S., £ 8.25. ISBN 0080139698

Pauli, W., General Principles of Quantum Mechanics, Berlin-Heidelberg-New York, Springer-Verlag 1980, XII, 212 S., DM 29,–. US $ 17.20. ISBN 3-540-09842-9. (Translation from the German)

Gokhfeld, D. A. / Cherniavsky, O. F., Limit Analysis of Structures at Thermal Cycling, Alphen aan den Rijn, Sijthoff & Noordhoff 1980, XXVIII, 537 S., ISBN 90-286-0455-3

Landau, L. D. / Lifschitz, E. M., Lehrbuch der Theoretischen Physik IV, Relativistische Quantentheorie **(Berestetzki, W. B. / Lifschitz, E. M. / Pitajewski, L. P.),** 4., bearb. Auflage, Berlin, Akademie-Verlag 1980, XII, 586 S., 24 Abb., DM 38,–. BN 7625470 (5436/4)

Serre, J.-P., Trees, Translation from the French, Berlin-Heidelberg-New York, Springer-Verlag 1980, IX, 142 S., DM 48,–, US $ 28.40, ISBN 3-540-10103-9

Frauenthal, J. C., Mathematical Modeling in Epidemiology, Universitext, Berlin-Heidelberg-New York, Springer-Verlag 1890, VII, 118 S., DM 36,–. US $ 21.30. ISBN 3-540-10328-7

NACHRICHTEN

Eighth International Conference on Numerical Methods in Fluid Dynamics

June 28—July 2, 1982

The Eigthh International Conference on Numerical Methods in Fluid Dynamics will be held in West Germany, June 21—25, 1982. The Organizing Committee is pleased to invite all interested scientists and engineers to attend.

The Conference will be held at the Rheinisch-Westfälische Technische Hochschule Aachen.

Contributed papers will be selected on the basis of a 1000-word abstract in English to be submitted before December 15, 1981.

Persons interested in receiving additional information on the conference should write to the

Conference chairman:

Professor Egon Krause, Ph. D.
Aerodynamisches Institut
Wüllnerstr. zw. 5 und 7
D-5100 Aachen, West Germany
Tel.: (0241) 805410

XI. International Symposium on Mathematical Programming

The Symposium will be held at the University of Bonn, August 23—27, 1982. It will be organized by the Institut för Ökonometrie und Operations Research and the Sonderforschungsbereich 21 (DFG) of the Rheinische Friedrich-Wilhelms-Universität Bonn.

Dates and Deadlines

Pre-registration should be applied for at earliest convenience but not later than October 1, 1981. — Deadline for contributed papers: April 1, 1982, the abstracts of the papers are due by June 1, 1982.

Mailing Address

Math. Progr. Secretariat
c/o Institut für Operations Research
Nassestraße 2
D-5300 Bonn 1, W. Germany
Telefon: (0228) 739203, Telex: 886657 unibo d

Achim Bachem
Martin Grötschel
Co-Chairmen
Organizing Committee

Bernhard Korte
Chairman
Program Committee

Berichtigung

O. Becker

Berichtigung zu meiner Arbeit „Zur Berechnung von Integralen über die Balkenfunktionen", ZAMM 61 (1981) 6, 271

Im Anhang A ist für den Fall $r = s$ bei den Integralen

$\int y_r' y_s''' \mathrm{d}x$, $\int y_r'' y_s'' \mathrm{d}x$, $\int y_r'' y_s''' \mathrm{d}x$, $\int y_r''' y_s' \mathrm{d}x$, $\int y_r''' y_s'' \mathrm{d}x$, $\int y_r''' y_s''' \mathrm{d}x$ $\lambda_s^4[\ldots]/N_{ss}$ an Stelle von $[\ldots]/N_{ss}$ zu setzen.